Maximizing Energy Savings and Minimizing Costs

John Studebaker, Ph.D.

Maximizing Energy Savings and Minimizing Costs

John Studebaker, Ph.D.

THE FAIRMONT PRESS, INC.

CRC Press
Taylor & Francis Group

Library of Congress Cataloging-in-Publication Data

Studebaker, John M.
Maximizing energy savings and minimizing costs / John Studebaker.
p. cm.
Includes index.
ISBN 0-88173-392-X (alk. paper) -- ISBN 0-88173-393-8 (electronic) -- ISBN 1-4200-8872-6 (taylor & francis distribution)
1. Electric utilities--Rates--United States. 2. Electric utilities--United States--Costs. 3. Gas companies--Rates--United States. 4. Gas industry--United States--Costs. 5. Energy consumption--United States.
I. Title.

HD9685.U5S7946 2008
333.793'230973--dc22

2008029737

Maximizing Energy Savings and Minimizing Costs / John Studebaker.

Published by The Fairmont Press, Inc.
700 Indian Trail
Lilburn, GA 30047
tel: 770-925-9388; fax: 770-381-9865
http://www.fairmontpress.com

Distributed by Taylor & Francis Ltd.
6000 Broken Sound Parkway NW, Suite 300
Boca Raton, FL 33487, USA
E-mail: orders@crcpress.com

Distributed by Taylor & Francis Ltd.
23-25 Blades Court
Deodar Road
London SW15 2NU, UK
E-mail: uk.tandf@thomsonpublishingservices.co.uk

Printed in the United States of America
10 9 8 7 6 5 4 3 2 1

0-88173-392-X (The Fairmont Press, Inc.)
1-4200-8872-6 (Taylor & Francis Ltd.)

Dedication

To Virginia

Table of Contents

Introduction

MAXIMIZING ENERGY SAVINGS AND MINIMIZING COSTS

There are so many different *(so-called)* energy *fixes* available, that many energy users are hesitant to do anything because of the apparent complexity of these *fixes*. Many large energy users have completely lost sight of the fact that they may not need *(at least initially)* time-consuming, large investment *fixes*. Strange as this may sound, many users have little or no practical knowledge about the basics of their energy purchases.

This book covers the basics of rate structures, how they apply to energy purchases, the components of the purchase process, and the methods and techniques required for maximizing energy savings and minimizing costs.

This book provides the foundation upon which any successful, long-term energy strategy must be based. The foundation of energy cost minimizing process requires the knowledge of *how current energy costs are developed and applied*. The foundation of a building is seldom seen or considered, but if it is defective the building will fail. It is the same for energy; if the *knowledge* foundation is lacking or defective, the entire process will fail, causing unnecessary costs to be present and potentially not even recognized.

Section I

Electricity

Chapter 1

Regulation of Utilities

UTILITIES—AN OVERVIEW

Electricity

Many changes are occurring in the electric utility industry. Deregulation of the electricity industry has progressed on a rather ununified state-by-state basis.

Electric utilities are becoming market responsive if they want to remain viable in their industries. The long-term outlook for electrical power alternatives is good and for many companies the savings realized through these various alternatives will be great. To be able to understand how changes occur in electric/natural gas utilities, a knowledge of what they are and how they are regulated must be available.

In electricity utility service, the commodity being sold generally is produced, at least in theory, by the utility that serves the customer. The process of the origination points of the commodity (electricity/natural gas) are shown on Figure 1-1.

Figure 1-1 illustrates that while electricity is a generated commodity, natural gas is a naturally occurring commodity. While this comparison may seem to be of no particular difference to the customer, it is of great importance to an electric utility. If an electric utility generates its commodity, it relies, at least in part, upon its retail customers to purchase the electricity it generates.

Another electric utility worry can be customer-distributed generation of electricity. In this scenario, the customer installs a generation system that provides both electrical as well as thermal (heat) outputs to supplement and reduce dependence upon utility supplied electricity.

Although distributed generation is not a new process, its use has become more common as utility-supplied electricity has become more costly. Many electricity customers currently have options to typical utility-supplied electricity. The question an electricity user has to answer is: which method of electricity purchasing is most reliable and cost effective?

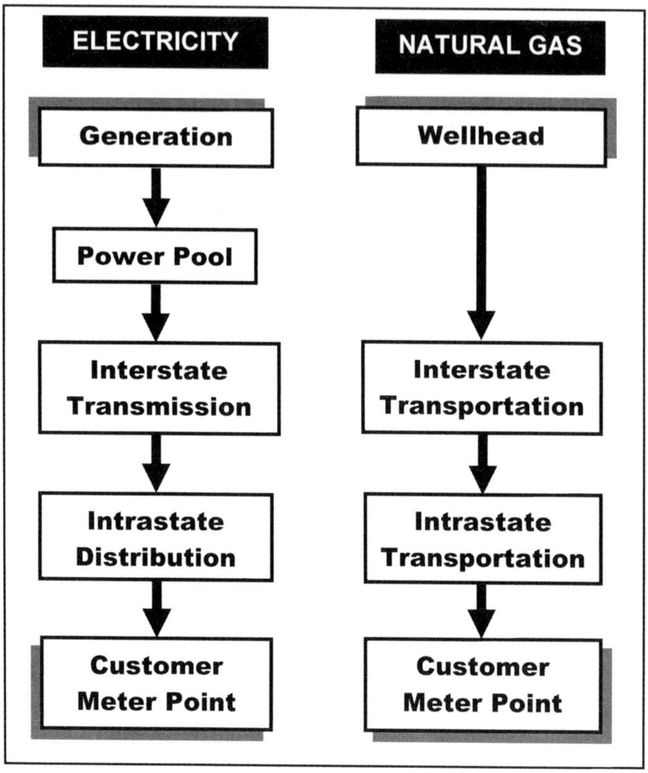

Figure 1-1. Commodity Origination Points—Electricity vs. Natural Gas

To answer this question, it is important to know what is available and how to most effectively utilize the best options. This publication will provide information needed by a customer to utilize the least expensive energy available while assuring that supply integrity is maintained.

THE REGULATION PROCESS IN GENERAL

The Regulation Process

Utilities, generally speaking, are considered to be regulated monopolies. A regulated monopoly is an entity that has a protected service territory and has virtually no competition since no other provider of the same commodity can compete for the protected service territory's customer.

Since a position of no competition can lead to abuse in both cost and service, a check and balance system is in place that requires utilities to be held accountable for costs, services, etc. Utilities are regulated in at least two areas, Interstate (between states) and Intrastate (within a state).

Interstate Regulation (Between States)

Interstate regulation in electricity and natural gas is the responsibility of the Federal Energy Regulatory Commission (FERC). This agency was created in 1977 and has the responsibility for oversight and regulation of interstate transportation policies and rates concerning electricity. Since this agency has these responsibilities, it would be well to understand its impact on rates and transportation conditions.

Electricity and natural gas distribution from one state to and for use in another state is always interstate (between states) and FERC has jurisdiction over the rules and regulations applying to it.

The Commission is composed of five members who are appointed by the President, with the advice and consent of the Senate. More and more the rulings of FERC meet with something less than enthusiasm on the part of the utilities it regulates, due primarily to its increasingly open access policies with relationship to utility competition.

Know what is happening at the federal regulatory level so that as changes occur, you will have a strategy in mind. The FERC website is a good source of information to be kept up to date on the subject of utility regulations—http://www.ferc.gov

Federal Entities

These types of utilities are of federal origin and regulation. Generally, intrastate regulation is not applicable since these types of utilities are operated on an interstate basis. The federal government has the overall responsibility of regulation and operating procedures.

As a rule, these utilities wholesale the majority of their power to for-profit investor-owned companies, municipalities, and cooperatives who in turn are regulated by their respective regulatory bodies. When direct sales are made to customers, regulation parameters are by federal guidelines.

The Influence Intrastate Regulation has on Utility Costs

State agencies regulate the intrastate distribution of electricity/natural gas. A utility typically uses a combination of inter (between states) and intra (within a state) components and therefore both federal and state regulation occurs.

Since state agencies provide predominate, day-to-day regulatory functions concerning electricity/natural gas, it is important to know that they function in a manner that benefits the customer.

One factor in intrastate regulation is the number of regulatory agencies that allow the utilities they regulate to not offer truly cost-of-service rate schedules. It does a customer no good to have interstate access to less expensive electricity/natural gas if the intrastate utility does not allow transportation of that electricity/natural gas through their individual systems.

Some Areas Regulated by FERC

1. Establishes and enforces rates and charges for electric/natural gas transmission and sales for resale. (*Very important to the deregulation of electricity.*)
2. Establishes and enforces operational characteristics, rates, and charges for electric/natural gas interconnections.
3. Certifies small power production and cogeneration facilities.
4. Issues and enforces licenses for nonfederal hydroelectric power facilities.
5. Issues and enforces certificates for construction and abandonment of interstate electricity transmission facilities.
6. Establishes and enforces rates and charges for distribution and sale of natural gas.
7. Establishes and enforces oil pipeline rates, charges, and valuation.
8. Establishes and enforces oil pipeline common carrier duties.
9. Hears appeals from Department of Energy remedial orders and denials of adjustments.

THREE TYPES OF INTRASTATE REGULATION
(Within State Boundaries)

1. For-profit Utilities

The state regulatory agencies usually take the form of public service commissions (PSC) or public utility commissions (PUC). The functions of these entities are to regulate the intrastate distribution and operation of utilities. These agencies also determine and approve individual utility rates of return, grant franchises to utilities for specific areas of operation, and in general, regulate the operation of utilities which are within a given state.

Although PSC or PUC structures are the most common forms of state regulation, other methods are used. In some states, these commissions regulate utilities only outside the incorporation limits of a municipality or city such as in the states of Georgia, Texas, etc.

Also, there are some strange situations that occur in a few states. For instance in the State of Texas, the Public Utility Commission regulates electricity outside of municipalities; however, outside of these same municipalities is an entity called the Railroad Commission of Texas regulates natural gas.

Generally, a retail customer will have more contact with state regulatory agencies than with federal agencies. Since state agencies determine the rate of return and approve or disallow rate increase requests of utilities, the likelihood of involvement with these agencies will be greater.

State Intervener Groups

To remain informed on utility matters, a knowledge of the operation and function of the state agencies is required. To follow state regulatory matters can be very time-consuming and costly if done on an individual basis. One alternative to this is to become a member of a state energy users group commonly called state intervener groups. These groups are comprised of numerous individuals that have common concerns—typically electricity/natural gas costs and regulations.

Generally, within each state there are groups of electricity/natural gas users that form formal or informal user groups to intervene in state utility rate cases. Many times these user groups can have a great impact upon pending utility regulatory matters. Often a well-organized cohesive group of utility customers can have a much greater impact on the outcome of a pending utility rate case than can an individual utility customer.

Collectively these groups can accomplish much at the state regulatory level. All user or intervener groups are required to register with the appropriate State Regulatory Agency. Therefore, to determine whether a state has intervener groups, contact the appropriate state regulatory agency. Also, check the Internet for intervener groups with websites.

These customer intervener groups may take the form of continuing well-organized entities or they may be as-needed ad hoc groups. Whatever the individual group's form may be, any utility user should at least investigate the groups that may be available in their particular state.

For a particular utility rate case, a utility user can contact the state utility regulatory agency for a list of the intervener groups that may be participating in that particular rate case. Legally, all intervener groups must register with the appropriate state regulatory agency before they can participate in any utility rate case.

Public Utility Commission Periodic Reports

Periodically, all public utility commissions that regulate for-profit utilities are required to issue status reports. These reports are generally issued on a weekly basis. The websites of these commissions provide data on all currently pending rate cases within their jurisdictions. If a public utility commission regulates the customer's utility, it is important to keep updated on what is going on at the commission level. Typically, the following information is detailed in these reports:

1. A docket or case number of the proceeding being discussed
2. A brief report concerning the status of the proceeding
3. Identification of the parties in the proceeding
4. Timetable and place of next commission meeting concerning the proceeding

State Enterprise Zones

Enterprise zones and economic development zones are present in many states. These zones are created to stimulate the economy in a given geographic area of a state and have nothing to do with utilities specifically. But in their effort to stimulate the economy, they generally allow credits for new or increased usage of utilities in the enterprise zone areas.

Typically, the utility incentives take the form of credits or concessions on new or increased utility usages with the credit or concession gradually disappearing over a period of time, generally 5 years.

These credits or concessions are directly deducted from monthly billing for the utility involved. Electricity and natural gas are both normally included. These credits or concessions do not cost the utilities since they are allowed a corresponding offset on their state tax liability. To determine whether you are located in an enterprise/economic development zone, contact your state government offices and request "enterprise/economic development zone information" including geographic boundaries and specific credits or concessions available.

Figure 1-2. US State Regulatory Agencies

Alabama Public Service Commission
www.psc.state.al.us

Regulatory Commission of Alaska
www.rca.alaska.gov

Arizona Corporation Commission
www.cc.state.az.us

Arkansas Public Service Commission
www.arkansas.gov/psc

California Public Utilities Commission
www.cpuc.ca.gov/puc

California Energy Commission
www.energy.ca.gov

Colorado Public Utilities Commission
www.dora.state.co.us/puc

Connecticut Department of Public Utility Control
www.state.ct.us/dpuc

Delaware Public Service Commission
www.delpsc.delaware.gov

District of Columbia Public Service Commission
www.dcpsc.org

Florida Public Service Commission
www.psc.state.fl.us

Georgia Public Service Commission
www.psc.state.ga.us

Hawaii Public Utilities Commission
www.puc.hawaii.gov

Idaho Public Utilities Commission
www.puc.state.id.us

Illinois Commerce Commission
www.icc.illinois.gov

Indiana Utility Regulatory Commission
www.in.gov/iurc

Iowa Utilities Board
www.state.ia.uslgovernment/com/util

Kansas Corporation Commission
www.kcc.state.ks.us

Kentucky Public Service Commission
www.psc.ky.gov

Louisiana Public Service Commission
www.lpsc.org

New Orleans City Council Utilities Regulatory Office
www.cityofno.com

Maine Public Utilities Commission
www.state.me.us/mpuc

Maryland Public Service Commission
www.psc.state.md.us/psc

Massachusetts Department of Telecommunications and Energy
www.mass.gov/doer

Michigan Public Service Commission
www.michigan.gov/mpsc

Minnesota Public Utilities Commission
www.puc.state.mn.us

Mississippi Public Service Commission
www.psc.state.ms.us

Missouri Public Service Commission
www.psc.mo.gov

Montana Public Service Commission
www.psc.mt.gov

Nebraska Public Service Commission
www.psc.state.ne.us

Nevada Public Utilities Commission
www.pucwebl.state.nv.us/pucn

New Hampshire Public Utilities Commission
www.puc.state.nh.us

(Continued)

Figure 1-2. US State Regulatory Agencies (*Continued*)

New Jersey Board of Public Utilities
www.state.nj.us/bpu

New Mexico Public Regulation Commission
www.nmprc.state.nm.us

New York Public Service Commission
www.dps.state.ny.us

North Carolina Utilities Commission
www.ncuc.commerce.state.nc.us

North Dakota Public Service Commission
www.psc.state.nd.us

Ohio Public Utilities Commission
www.puc.ohio.gov

Oklahoma Corporation Commission
www.ccc.state.ok.us

Oregon Public Utility Commission
www.puc.state.or.us

Pennsylvania Public Utility Commission
www.puc.state.pa.us

Rhode Island Public Utilities Commission
www.ripuc.org

South Carolina Public Service Commission
www.psc.sc.gov

South Dakota Public Utilities Commission
www.puc.sd.gov

Tennessee Regulatory Authority
www.tennessee.gov/tra

Tennessee Valley Authority
www.tva.gov

Texas Public Utility Commission
www.puc.state.tx.us

Texas Railroad Commission
www.rrc.state.tx.us

Utah Public Service Commission
www.psc.state.ut.us

Vermont Public Service Board
www.state.vt.us/psb

Virginia State Corporation Commission
www.state.va.us/scc

Washington Utilities and Transportation Commission
www.wutc.wa.gov

West Virginia Public Service Commission
www.psc.state.wv.us

Wisconsin Public Service Commission
www.psc.state.wi.us

Wyoming Public Service Commission
www.psc.state.wy.us

Overview

For-profit (Investor-owned) Utility Regulation

- Intrastate regulatory body—state regulatory agency
- Utilities in the classification—Any utility that is in business for the stated purpose of making a profit and is owned by investors through the purchase shares of stock.

- Regulation process—Any rate change requires at least the following items:
  - A public notification of intent to change a rate.
  - Adequate notification period prior to actual rate case presentation (this period defined by state law) to allow interested parties to study the merits of the change request.
  - Presentation of rate requests at a hearing open to the public before the appropriate regulatory agency.
  - Allow input from interested customer groups (interveners) relating to rate change requests. The state regulatory body, based upon testimony presented at rate case hearings, approves actual rates that will be put into effect.

2. Municipal Utilities

When municipalities regulate utilities they are, as a rule, self-governing—that is, they are not subject to state regulatory rulings. Generally, when municipalities undertake the providing and regulation of utilities, they are purchasing the commodities at wholesale rates from a for-profit or federally regulated utility, and then retailing these utilities to the public. This is especially true with electricity and natural gas. In the case of water and sewage, the municipality usually has control or jurisdiction over the entire process.

Municipal utilities are generally presided over by a utility commission or board of appointed or elected members. As in the case with all utility regulatory agencies, *due process* must occur before changes can be made with respect to rates and conditions under which utilities are provided.

This means that public notice must be given and adequate time allowed for public input prior to a change being instituted. Typically, municipal utilities do not have as many rate classes or options as do for-profit investor-owned utilities since they generally do not have as diverse a class of customers.

Overview

Municipal Utility Regulation

- *Regulatory body*—Generally, self-regulated by the body or agency selected by the municipality to oversee utility matters; can be separate for each utility regulated.

- *Utilities in this classification*—Any utility operated and regulated by a municipality on a not-for-profit basis.

- *Regulation Process*—Similar to for-profit investor-owned utilities
  - — Public notification
  - — Adequate notification period
  - — Presentation of rate request
  - — Allow input from interested customer groups

3. Rural Electric Cooperative Utilities

Cooperative utilities are formed generally when a for-profit investor-owned utility elects not to serve a geographic area or customer base. Cooperative utilities usually serve rural areas where there is not a large customer load base. Generally, power is purchased at wholesale from a for-profit utility and distributed by the cooperatives' lines and/or pipes to the individual customers' locations.

Cooperative utilities are like municipal utilities in that they are self-regulated but also are required to provide *due process* before instituting changes in the utility rate base. They are also very different from any other type of utility since they are classified as a "cooperative" entity.

The term *cooperative* as far as utilities are concerned, literally means that each customer is a part owner of the utility and as such, at least in theory, has their proportionate say in how the utility is operated.

They are similar to all other types of utilities in rate change cases. Also, they propose rate changes, hear customer input, and all other *due process* practices before actually instituting rate changes. Cooperative utilities, as compared to municipal utilities, are smaller in terms of total energy supplied, and in general, are located in rural types of service areas.

Overview

Cooperative Utility Regulation

- *Intrastate regulatory body*—Generally self-regulated. All customers are part owners of the utility and as such have their proportionate say or vote based upon their usage in relation to other customers. In practice many times, a board of overseers is appointed to represent customer interests with respect to the utility operation.

- *Utilities in this classification*—Any utility defined as cooperative and that is owned and operated by the customer served.

- *Regulation Process*—Similar to other types of utilities and includes:
  - Public notification
  - Adequate notification
  - Presentation of rate request
  - Allow input from customers

SYNOPSIS—REGULATION OF UTILITIES

Outlined in this chapter are the basics of the utility process. If an effective utility cost reduction program is to be put into place, an understanding of how the utility process works, is essential to the success of the program.

Chapter 2

Developing a Strategy for Reducing Electricity Costs

THE COST REDUCTION STRATEGY

The first step in this undertaking is the obtaining of accurate information relating to the subject being investigated. The first step in understanding begins with the obtaining of data that relates to the pricing on a unit basis of the commodity being purchased. Since all utilities are regulated in one way or another, written records of usage and pricing data, as it relates to customers, must be available to any interested party that requires it.

Items Needed from the Utility and State Regulatory Agency

Listed following are the mandatory basic informational items that must be obtained before any understanding of utility rates will be realized. All of these items are a matter of public record and must be made available to anyone who requests them. They are typically available from at least two sources: (1) the serving utility and (2) the regulatory agencies.

Generally, if a request for information is received from a customer of the utility, there generally is no problem or cost involved. All utility rate information that is approved by a regulatory body for use in determining rates and conditions to which a customer is subject, must be a matter of public record, and as such, available for public inspection.

Typically, the utility service representative responsible for the customer involved is contacted and a request is made for the information needed. The importance of obtaining the information illustrated in Figure 2-1 cannot be overstated since these items are basic to understanding utility costs.

With relation to information that is required from the state, the best way to proceed is to contact the state agency involved. As a rule, there should be no problem in obtaining either sales tax or economic development/enterprise zone information from the state.

In Figure 2-1 is shown the various items needed prior to analyzing any electricity billing. This figure lists the items that are normally obtained from the service representative of the utility involved to provide a thorough and comprehensive analysis.

EVALUATE BASIC ELECTRIC DATA

Complete Rate Schedule

A complete rate schedule covers all rates, terms and conditions that are approved in a rate case. All classes of customers are addressed-residential, commercial and industrial. These schedules can range from several sheets to several hundred sheets in length.

Contained in this information will be all data relating to customer rates, costs, terms for service, etc. The importance of this source docu-

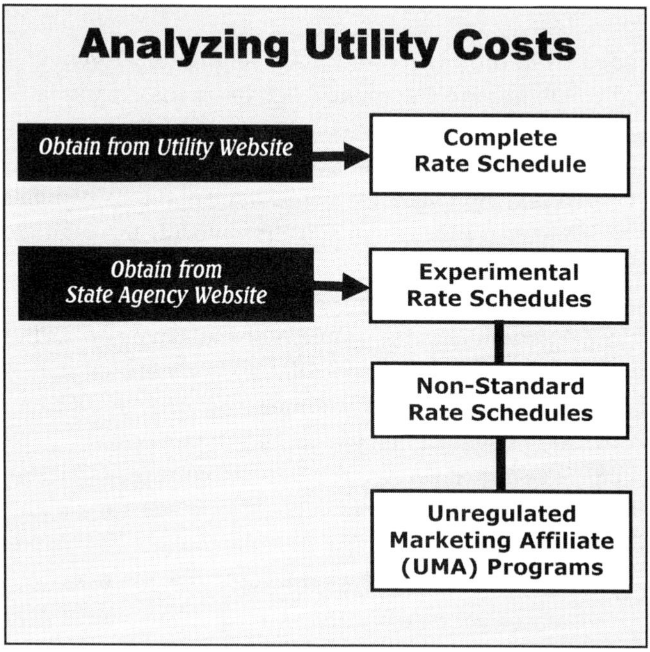

Figure 2-1. Basic Electric Utility Data Needed

ment cannot be overemphasized since it is absolutely mandatory for an understanding of utility costs. Make certain that the request is made for a complete rate schedule since utilities tend to provide only the particular schedule that currently applies to the customer making the request.

The availability of a complete schedule cannot be over emphasized since only then can comparisons be made between different rates and options. A typical complete rate schedule will contain the following items:

1. Complete list and explanation of all customer rates available.

2. Complete list of all items or riders that modify or change rate costs.

3. Alternative rates that may be available on a "customer request" basis for certain customer classes.

4. Information on "special" rates that may be available in certain circumstances.

5. Complete explanation as to how all cost components of utility usage are measured and applied. Complete rate schedules remain in effect until a new rate case is filed and approved by the appropriate regulatory agency. Only one complete schedule is required for a given utility since all customer classes are addressed therein.

Experimental Rate Schedules

Experimental rates are not normally contained in complete rate schedules since they are developed on an experimental basis by utilities and are not mandated for any particular customer. Those types of rates are not available from all utilities, but if they are, they can be a source of cost reduction potential.

Experimental rates are developed by the utility and approved on an experimental basis by the applicable regulatory agency. The experimental category allows the utility to evaluate the potential for a different type of rate structure.

Since these rates are never mandated, they are used only on a customer voluntary basis. If the customer chooses an experimental rate and it results in an increased cost, then utility may not assess any charge higher than what would have resulted from the regular schedule of rates.

If an experimental rate proves successful, the typical next step is to include it as an optional rate in the base rate schedule and not mandatory for any customer class. Since the final step is to change the *optional* classification to a *mandatory* category in the base rate schedule, it remains important to keep up-to-date on experimental rates since long-term they have a way of becoming mandatory for some customer classes.

The most common experimental rate structure currently being used in electricity is the real time pricing (RTP) structure. If RTP structures are available, evaluate them to determine the immediate applicability as well as the long-term implications in case they are later included in the base rate schedules as mandatory for your customer class.

Steps in Establishing Experimental Rates

Step 1. Experimental Rate
Step 2. Optional Rate
Step 3. Mandatory Rate

An example of what an experimental real time pricing electric could look like is shown in Figure 2-2 following.

Analysis of Figure 2-2

This experimental rate allows a customer to purchase electricity on an hour-by-hour basis based upon (kWh) usage only. The problem with a rate of this type is that usage (kWh) charges will probably be more expensive at times when the customer usage will be the greatest.

For example, in Figure 2-3—Example of a Daily Usage Cost Printout, the maximum charge occurs from 10:00 a.m. through 7:00 p.m. on normal workdays (Monday-Friday). In this example, very costly electricity will be consumed during most customers' largest usage periods.

Generally, this type of rate will be of benefit only to the customer that can shift electrical usage to the utility's normal off-peak periods, which would usually be during the evening, night, early morning, weekends and holidays.

Most electricity users do not have the flexibility to shift usages to the extent needed to benefit from this type of rate. However, if the customer can alter the usage patterns on a daily/hourly basis, a rate of this type can be very cost effective. An item-by-item analysis of this rate follows:

1. *Schedule*

   This schedule number (EX-RPT-7) designates the rate case identifica-

Experimental Rate Schedule

1. **Schedule**: EX-RPT-7—(Experimental Real Tune Pricing)

2. **Applicability**: Applicable to any general service customer with service delivered at a voltage level of 4,160 or greater. To qualify for this rate, the customer must have maintained a monthly billing demand level of at least 750 kW for the last 12-month period.
This rate is available on an experimental basis and the minimum term is for a one-year period.

3. **Base Monthly Rates**:
   A. Customer Charge $700.00
   B. Demand Charge—per incremental kW (None)
   C. Energy Charge:
      (1) Base Energy Charge-per kWh (Fuel Cost) $.0126
      (2) Energy Charge-per kWh (Variable) ($.0216-.6130)

4. **Cost Data**: Cost data shall be calculated on an hourly basis every day. No later than 8:00 a.m. every day the next day's hourly rates beginning at 12:01 a.m. will be provided to the customer via a telecommunication link with the customer mandatory dedicated telephone line. This data will provide the (24) hourly variable energy charge components for the next billing period. This data will be provided every day of the year. The energy charge components will reflect the utility's actual costs as detailed in the Public Utility Commission approved rate filing 416 dated 07-01, pages 143 through 156.

5. **Variable Energy Charges**: Variable energy charges will range from a minimum of $.0216/kWh to a maximum of $.6130/kWh per hourly measurement period.

6. **Minimum billing**: The minimum monthly billing shall consist of the following item:
   A. Customer charge $700.00

7. **Miscellaneous Provisions**:
   A. There shall be no demand (kW) charges applied to this rate.
   B. The utility will provide and maintain the appropriate metering and related equipment to accurately measure the kWh consumption on an hourly basis, at no cost to the customer.
   C. The utility retains the right to limit the number of customers on this rate.
   D. The utility retains the fight to withdraw this rate upon one month's notice.

Figure 2-2.
Example of an Electricity Experimental Real Time Rate Schedule

tion number assigned to this rate by the utility/regulatory agency as a result of the utility's case concerning this rate. If a customer wanted to examine all pertinent data presented in this rate case filing, this could be accomplished by requesting the data from either the utility or the regulatory agency by schedule number (EX-RPT7).

2. *Applicability*

This section addresses the type of customer that can be served on this rate. In this particular rate, the customer must be served at a primary voltage level of from 4,160 or greater. Also, the customer must have maintained a minimum billing demand level of 700 kW monthly for the last 12 months.

Generally, applicability provisions are instituted because the utility has determined that this minimum voltage/demand threshold would be required by a customer to benefit from the rate. Also sometimes, especially on a rate of this type (experimental), the utility may want to test the rate's validity or applicability only to a certain type or class of customers.

3 *Base Monthly Rates*

A. Customer Charge

This is the minimum monthly charge that an individual customer would have to pay to be served on this rate. This charge covers the utility's cost of maintaining and reading the meter and miscellaneous other monthly billing cost items.

B. Demand Charge

This particular type of rate has no defined demand charge. Nevertheless, demand costs are calculated and included in the energy charges associated with this rate. Although a rate with no demand (kW) charge may seem to be a very cost effective type of rate, the truth of the matter is that demand costs are calculated and included in the energy charge portion of this rate.

In fact, a rate of this type actually would probably be more expensive for most customers that could not shift usage patterns on a daily/hourly basis, which is very difficult to do. Just because demand is not included as a specific billing item does not mean that its impact on the utility's costs has not been considered. It has, and in this particular type of rate, is included in the variable energy charge portion of the billing.

C. Energy Charge:

(1) Base Energy Charge-per kWh (Fuel Cost)

This energy charge of $.0216/kWh represents the utility's fuel cost to operate their generation equipment. This charge is sometimes called fuel cost adjustment and it is always applied to the energy (usage, kWh) portion of the billing. All customers of a utility are assessed the same charge on each kWh used. All of these charges are approved by the appropriate regulatory agency.

(2) Energy Charge-per kWh-(Variable)
This charge represents the variable usage (kWh) charges by the utility on a daily/hourly basis. The extremely wide range of variability, over \$.5914 per kWh (\$.6130-\$.0216 = \$.5914) is due to the cost of electricity that is experienced by the utility based upon its generation load/utilization at various times of the day. Since this rate is variable and is priced by the day and hour, it actually can change on a daily basis. Figure 2-3 illustrates what a daily usage cost printout from the utility might look like.

In this hourly kWh data is the cost of electricity during the (24) hours of this particular day (Wednesday); the kWh cost ranges from \$.0216 (2:01 a.m./3:00 a.m.) to \$.1910 (1:01 p.m./2:00 p.m.) with the higher costs being between 6:01 a.m. and 10:00 p.m. If a customer uses the majority of the electricity during these hours, then the cost for the electricity will be very high.

Rate—EX-RTP-7 LOW **Data for: <u>Wednesday</u>** HIGH

Time of Day	Hourly kWh Cost	Time of Day	Hourly kWh Cost
12:01am— 1:00am	$.0312	12:01pm— 1:00pm	$.1317
1:01am— 2:00am	$.0361	1:01pm— 2:00pm	**$.1910**
2:01am— 3:00am	**$.0216**	2:01pm— 3:00pm	$.1833
3:01am— 4:00am	$.0297	3:01pm— 4:00pm	$.1761
4:01am— 5:00am	$.0361	4:01pm— 5:00pm	$.1823
5:01am— 6:00am	$.0361	5:01pm— 6:00pm	$.1310
6:01am— 7:00am	$.0476	6:01pm— 7:00pm	$.1107
7:01am— 8:00am	$.0621	7:01pm— 8:00pm	$.0916
8:01am— 9:00am	$.0791	8:01pm— 9:00pm	$.0888
9:01am— 10:00am	$.0816	9:01pm— 10:00pm	$.0611
10:01am— 11:00am	$.1273	10:01pm— 11:00pm	$.0426
11:01am— 12:00pm	$.1197	11:01pm— 12:00am	$.0371

Figure 2-3. Example of a Daily Usage Cost Printout

If however, a customer has the majority of the usage between the hours of 11:01 p.m.-7:00 a.m., then the cost per kWh would range from $.0312 (12:01 a.m./1:00 a.m.) to $.0476 (6:01 a.m./1:00 a.m.).

Depending upon usage patterns, a real time pricing structure could result in savings when compared to a typical rate schedule for this same type of usage pattern.

Basically a real time pricing rate is generally only cost effective if a customer can vary their electrical loads on a daily time related basis; or, if they ordinarily operate during periods of low utility load demand intervals of time-typically evening, night, early morning and weekends. If a customer was served on this rate, they would receive a daily printout of data similar to the one shown here for every day of the month, including weekends and holidays.

4. *Cost Data*

   This section explains how the hourly kWh charges will be calculated.

5. *Variable Energy Charges*

   This section simply restates the information shown in item 3.C.(2) Energy Charge-per kWh-(Variable).

6. *Minimum Billing*

   This section details what will constitute a minimum monthly customer bill. Basically, the minimum charge will consist of the customer charge ($700).

7. *Miscellaneous Provisions*

   This section enumerates miscellaneous provisions associated with this experiment rate schedule. As can be seen from analyzing the provisions of this example of a real time pricing rate, it is not a rate for all customers.

   Generally, a rate of this type will not be cost effective for customers who operate on a one-shift basis during normal first-shift work. However, if a customer operates hours other than these, or can shift their usage on a daily and hourly basis, or can work many hours on weekends *(Saturday and Sunday)*, then a rate of this type may prove to be very cost effective.

   Before changing to a rate of this type, always have the utility do an analysis of the past year for the facility/operation in question to determine the cost effectiveness of a change.

Non-Standard Rate Schedule

The non-standard rate schedule differs from the standard applicable rate schedule, as well as an experimental rate schedule, in the way it is developed and applied. Non-standard rate schedules are negotiated generally between a utility and a specific customer. Initially, they are generally discriminatory in nature and typically apply to larger customers. Rates of this type are negotiated and must be approved by the appropriate regulatory agency.

As a rule, non-standard rate schedules are developed for large user customers that have either extremely large loads or unusual use characteristics that are not addressed adequately in base rate structures. Most non-standard rate schedules occur in larger utilities with diverse customer bases.

To establish a non-standard rate schedule is at best a drawn out procedure. First, the utility, as well as the regulatory agency, has to be convinced of the need for a rate of this type—no easy task in itself.

Second, other utility customers may argue that this is a discriminatory rate since it could impact their utility costs unfavorable; and, that someone could have to pay for lost revenue that generally results from a non-standard rate schedule.

Non-standard rate schedules in the past have not been widely used or applicable to a large user bases, but this is changing with the potential of electricity deregulation becoming a reality.

Evaluate the complete rate schedule in terms of usage characteristics, and if large differences appear between actual usage patterns and those specified in the complete rate schedule, the potential for a non-standard rate schedule may exist.

To determine whether non-standard rate schedules are available, contact the appropriate utility regulatory agency and request a copy of any non-standard rate schedules that are currently available. An example of what a non-standard rate schedule might look like is shown in Figure 2-4.

Analysis of Figure 2-4

This example of a competitive service rider represents an electric utility's process to reduce specific customer's electricity costs when re-

Availability: Available at Utility's discretion to Commercial and Industrial Customers that have electric service requirements which are subject to effective competition. Effective competition exists if a Customer is located in Utility's service territory and has the ability to obtain its energy requirements from an energy supplier not rate regulated by the State Regulatory Agency.

Rate: Standard service rate provisions apply except the level of the demand and/or energy charges may be decreased for each Customer based on a consideration of Customer's load characteristics and lowest cost competitive energy supply.

Terms and Conditions of Service:

1. Customer must provide Utility with information that documents that Customer is not likely to take service provided by any other electric rate available from Company.

2. Minimum load served under this Rider is 1,000 kW.

3. Customer must execute an electric service agreement with Utility that will include:
   A. The minimum rate under this Rider, which will recover at least the incremental cost of providing service, including the cost of incremental capacity that is to be added while the rate is in effect.

   B. The maximum possible rate reduction under this Rider, which will not exceed the difference between the standard rate and the cost to the Customer of the lowest cost competitive energy supply.

   C. The term of service under this Rider—must be no less than three (3) years.

   D. The size of the load served under this Rider, which must be 1,000 kW at a minimum.

   E. An annual minimum charge to fully recover distribution costs.

Figure 2-4. Example of an Electricity Competitive Service Rider

quired to preserve or retain their load. This type of rate is becoming more common with the introduction of electricity deregulation. A serving utility could utilize a rate of this type to effectively retain a customer that could leave the utility's generation base, which could increase the utility's incremental base generation costs.

When a rate of this type is utilized, any cost reduction related revenue shortfall cannot be passed through to any other customer in any subsequent rate case proceeding. In the case of for-profit utilities, any revenue shortfall must be absorbed either through operational efficiencies or by the shareholders.

1. *Availability*

   A rate of this type is inherently limited in scope since the utility can, within this rider's written boundaries, choose which, if any, individual customers it will offer this ratemaking process to. Any utility shortfalls caused by revenue discounts offered through this type of ratemaking process normally cannot be passed through to any other customer.

   Also, the offering of this type of discount can preserve generation base load which can positively impact other customer classes by potentially reducing utility base generation costs which ultimately affects overall utility costs.

2. *Rate*

   This provision defines the rate or cost considerations applicable to this rate rider. Note that there are no definitive numbers in this section since the actual costs are determined with relation to each customer's worth to the utility.

3. *Terms and Conditions of Service*

   This section of the rider defines the terms and conditions under which the customer will be served.

   (1) *Customer responsibility* to *provide data* to *the utility.* This rider stipulates that the customer must verify, to the utility's satisfaction, that the customer will actively seek alternative suppliers if a discount is not provided.

   (2) *Minimum load.* In this case, the minimum load that the utility will consider for a discount is 1, 000 kW. If a customer did not have a single meter point demand of 1,000 kW, but instead had several meter points with each less than 1,000 kW; and, that these points were in the aggregate total of more than 1,000 kW, then a rider of this type might be utilized if the utility could be convinced that it was in their best interest. Potential customers that might fall in this type of situation could be fast food restaurants, small commercial multiple location establishments, or any multi-meter customer with an aggregate total of over 1,000 kW for all meter points in the serving utility's service territory.

   (3) *Customer/utility agreement.* In order to take advantage of this rate rider, the customer must execute *(sign)* an agreement with the utility that sufficiently addresses the following five categories:

A. *Minimum rate.* In rate riders like this, the utility is going to limit their "lost" revenue exposure by establishing limits on the rate discount that could be available to any customer. This particular provision states that after any discounting to the customer, the utility will at least receive revenue, sufficient to cover the basic cost of providing service including incremental capacity charge and any distribution cost related to the rate.

What this provision addresses is: generally no utility will sell electricity at a cost lower than their base cost of providing that electricity. The incremental capacity and distribution costs addressed here are those basic utility costs that represent the minimum values that the utility must recover on any rate discount they offer. If they were to discount further, the revenue they would receive would be less than their actual costs. The costs addressed in this section are generally on record at either or both state and federal regulatory agencies.

B. *Maximum rate reduction.* Not withstanding section *(A)* previously or section *(E)* following, the utility is not going to offer any customer a discount larger than they have to in order to retain that customer. This section states that the customer will probably have to demonstrate or document what competitive offers they have received in order for the utility to make a competing offer.

This type of provision can cause problems for the utility customer in confidentiality agreements they may have with potential non-utility providers. Sometimes a utility customer can simply execute a request for proposal (RFP) to alternative electricity providers—unregulated marketing affiliates (UMA), to convince the serving utility that the customer is serious about reducing their electricity costs.

As electricity is deregulated, more opportunity for arrangements like this will be available. Remember, trying a strategy like this has little downside but not trying can result in more expensive electricity costs as a result of not taking advantage of the available discounts in a given market area.

C. *Term* of *service.* The minimum term for an arrangement negotiated under the terms of this agreement will not be less than (3) years. Terms for agreements of this type generally range between 3-1 a years in length. A shorter, rather than longer term is better since changes can and will occur with both the customer as well as the utility that could make any long-term agreement look not so good when viewed in retrospect. Generally, any contract terms longer than (5) years should be seriously considered before implementation.

D. *Size* of *load (kW) served.* The minimum load (kW) that the utility will consider for discounts in this rider is 1000 kW, but there is no maximum kW limit. Items that should be considered by the customer when considering this section are as follows:
   (1) Is the 1,000 kW minimum through (1) meter or can meters of less than 1,000 kW each be combined to accumulate the 1, 000 kW minimum?
   (2) During the contract term, can the customer add to or take away from the kW quantity that was present at contract inception?

Utility Unregulated Marketing Affiliate Programs (UMA)

Some electric utilities currently have peak power demand (kVA/kWh) deficits. This means that even though a utility may not have a base load (kWh) problem, they may experience a generation capacity shortfall during some periods of a 24-hour day.

The utility can do several things to compensate for this generation capacity shortfall. They can construct new generation plants (supply-side planning) that are very expensive or they can offer their customers financial incentives to reduce demand during the utility's generation shortfall periods (demand-side planning).

Many utilities, through their UMA services, offer programs that encourage customers to reduce their demand needs by paying for or providing incentives for the particular items that favorably impact the utility's demand shortfall problems.

These programs range from *not worth much* to *extremely beneficial.* They change frequently and sometimes a specific amount of money is allocated for a program, which means that when the money is gone, the program is ended.

If a UMA program is available, there will generally be an in-house specialist that can be utilized for an onsite facility evaluation to determine the applicability of the program to a particular situation. The utility service representative can provide program information as well as arrange for an onsite evaluation by the appropriate specialist. These programs typically include the following items/processes although not all items are included in all programs.

1. Utility audits for rebate applicability
2. Fluorescent lighting
3. High intensity discharge lamping
4. Electronic ballasts
5. Efficient magnetic hybrid ballasts
6. Reflectors
7. Occupancy sensors
8. Miscellaneous lighting controls
9. Rooftop air conditioning
10. Window air conditioning
11. Electric chillers
12. Gas-fired air conditioning
13. Heat pumps
14. Boiler/water heaters
15. Cool storage-thermal storage
16. Energy management systems (EMS)
17. Energy efficient motor drives
18. Power factor correction capacitors
19. Thermal insulation and window film
20. Custom programs structured to individual customer requirements

(These programs are individually negotiated on a *customer/utility basis.)*

SYNOPSIS—DEVELOPING A STRATEGY FOR REDUCING UTILITY COSTS

Who is Responsible for Reducing Utility Costs?

The Federal Regulatory Commission (together with all state regulatory agencies) agrees that *the customer is ultimately responsible* for being on the most cost-effective rate. There are no state statutes that require a utility to ensure that a customer is served under the most economical or least costly rate available. Neither are there requirements that a utility refund any excess monies paid by a customer if they are on a correct rate, even if it is not the most cost effective. Since this is true, a customer must understand the utility cost reduction process to be able to determine whether the costs are as low as possible.

Due to the vast number of customers most utilities serve, and the changes and revisions their customers are constantly experiencing, it would be impossible to assure that any customer is at all times on the most economical rate available. Each customer is responsible for the cost effectiveness of the rate under which they are served. Unfortunately, many utility customers are unaware of the rate that forms the basis of their utility billing.

Through information provided in this publication, the correct approach to reducing utility costs can be understood. With this knowledge, the customer can determine the steps needed to be assured their utility costs are what they should be.

Chapter 3

Analyzing the Electric Utility Bill

ELECTRIC UTILITY COSTS

Analyzing the Electric Utility Bill

Since practically all electric utilities utilize different styles of billing formats, as well as types of information displayed on billings, there is no single "fits all' example. In view of the fact that there is not a single example that would apply to everyone's billing design, the following Figure 3-1. *Composite Electric Utility Billing Example* is shown.

While this example might be somewhat different from any actual utility billing, it does contain all major categories of data that would normally be on any actual billing. Once this example and its thirty separate items are examined and understood, any actual billing will be much easier to understand and decipher.

Explanation of Figure 3-1 Composite Electric Utility Billing Example

ITEM #1—Meter Number

This section shows the customer's individual meter number together with the beginning and ending meter readings for electricity usage (kWh) and demand (kW). The number shown in this section of the billing should match the number on the physical customer meter.

If the billing meter number and the physical meter number do not match, there is a problem. This discrepancy will cause an error in the electricity billing since the meter number has a direct correlation to Item #3 (Meter Constant). Always investigate the accuracy of the billing meter number since it is imperative to an accurate electricity cost.

ITEM #2—Beginning Reading (Usage-kWh)

The beginning figure indicates the beginning numeric meter reading when the previous month's meter reading was taken. The previous

BILLING DETAIL

Electric Billing: 01-01 — 02-01 (31 days)

1	2	3	4	5	6 ······ 7	8
Meter Number	Begin Reading	End Reading	Difference	Meter Constant	Consumption	Unit
7461187	02252	02717	465	1,200	558,000	kWh
7461187	0	1.17	1.17	1,200	1,404	kW

9	Rate	GSP-3	
10	Voltage Level	12.2 kVA	
11	Power Factor	71%	
12	Billing Demand	1,614 kW *(from 10-07 billing)*	
13	Contract Demand	1,200 kW *(from Contract 03-01)*	
14	**BILLING SUMMARY**		
15	Customer Charge:		210.00
16	Energy Charge:		
	(1st 200 x 1,614 kW	= 322,800 kWh x $.0513 /kWh)	16,559.64
	(2nd 100 x 1,614 kW	= 161,400 kWh x $.0427 /kWh)	6,891.78 =$24,912.66
	(Remainder	= 73,800 kWh x $.0198 /kWh)	1,461.24
17	Demand Charge:	(1,614 kW x $10.00 /kW)	16,140.00
18	Fuel Cost Adjustment	$.0023 ($.023 x 558,000 kWh)	1,283.40
19	Power Factor Adjustment:	($.7000 x 1,614 kW)	1,129.80
20	Facility Charge:		1,147.23
21	Facility Lease:		968.17
22	Special PUC Charge:	$.0011 ($.0011 x 558,000 kWh)	613.80
23	Nuclear Regulatory Cost:	(558,000 kWh x $.0006)	279.00
24	School Tax (3%)	(3%) on ($46,689.06)	1,400.00
25	Franchise Fee (1.5%)	(1.5%) on ($48,084.58)	968.17
26	State Sales Tax (7%)	(7%) (on $48,805.85)	3,416.41
27	TOTAL CURRENT BILL		$52,222.26
28	Previous Amount Due		$49,176.83
29	Payment Received 12/15		$49,176.83
30	TOTAL AMOUNT DUE 02/20		$52,222.26

Figure 3-1. Composite Electric Utility Billing Example

month's ending reading and this month's beginning reading should be the same reading.

Think of the kWh meter reading as similar to a non-resettable automobile odometer reading. When a replacement meter is installed in a customer location and its kWh register will be set to '0'. The entire time that the meter is in service at the customer's location, the kWh register will continue to read on an accumulating basis with no resetting to '0'. An actual month's kWh usage will be calculated by subtracting last month's ending reading (the billing month's beginning reading) from the billing month's ending reading. (For the sample month shown—02717-02252.) Remember, this total represents only the raw meter reading data. This reading total (465) multiplied by the meter constant (see Item #5) is utilized to arrive at the actual month's kWh usage.

ITEM #3—Ending Reading (Demand-kW)

The kW register is different from the kWh register in the respect that it measures maximum demand requirements averaged over either 15- or 30-minute periods throughout the month. It does not measure electricity usage (kWh). Also the kW register is reset to 10′ each month unlike the kWh register that reads continuously throughout its location at a specific customer facility. KW (demand) represents the peak electricity requirements of the customer during a 15- or 30-minute time period. Demand has *no* relationship to the total electricity usage (kWh) during the billing month. Think of kW and kWh in electricity as being comparable to water pressure (kW) and water flow (kWh).

ITEM #4—Difference

Both of the kWh and kW meter register beginning readings are subtracted from the ending meter register readings [kWh—(02717 – 02252) = 465 units} [kW—(1.17 – 0) = 1.17 units). These meter-register readings represent only the raw meter data; they do not represent the actual usage (kWh) or demand (kW) totals for the billing period.

ITEM #5—Meter Constant

Electricity meters typically do not measure the entire electricity flow that passes through. Rather on a continuous basis, meters measure a representative sample that is indicated by the meter constant—in this sample (1,200). The meter-multiplying constant is a meter calculation, generally determined by meter current transformer (CT) winding ratios to actual current flow through the meter.

Meter constants can range from (1) to (10,000) or greater. If the meter constant is correctly calculated, a small meter constant is no more accurate than a large meter constant. Generally, meters are calibrated to at least ± 1/10% (.01) over the full meter recording range. Electric meters generally do not measure actual current flow since smaller, less expensive meters can be utilized by incorporating a constant factor and having the meter actually measure a portion of actual current flow. In this sample, the meter constant is (1,200).

ITEMS #6 & #7—Consumption (kWh/kW)

Total consumption/demand is simply the calculation of the meter difference readings for kWh and kW registers multiplied by the meter constant. For this sample, these calculations are as follows:

kWh	02717 – 02252 = 465 × 1,200 = 558,000 kWh
kW	1.17 – 0 = 1.17 × 1,200 = 1,404 kW

ITEM #8—Units

Here are shown the units used to calculate the electricity usage data (kWh and kW).

ITEM #9—Rate Classification

The rate classification is the electric utility rate schedule identification that applies to the utility rate designation for the electricity allocated for the meter being utilized. The complete description of any given rate classification for electricity natural gas can be obtained from either the specific serving utility or the applicable regulatory agency involved. In this sample, the rate classification is GSP-3. The actual classification nomenclature (GSP-3) may/may not have any particular meaning depending upon the specific utility involved. In this particular sample, (GSP-3) means general service primary, Class 3.

ITEM #10—Voltage Level

This item indicates voltage level (12.2 kVA or 12,200 volts) the customer requires the utility to provide to their meter point. Generally, the higher the voltage level at the customer meter point, the less expensive will be the customer's incremental electricity (kWh/kW) cost.

The reason for this is, in general terms, the higher the voltage level at the customer meter point, the fewer will be the voltage step down transformation steps that will be required that will result in lower utility costs. In general, these reduced costs will be passed on to the customer. Based upon the customer's usage characteristics, there are practical limits to the voltage level at which they can be served at their meter point.

ITEM #11—Power Factor

This item indicates the efficiency at which the customer uses their electricity. Power factor is a measurement of the difference between two electricity measurements—kVA and kW. (kVA is 1,000 volt-amps; kW is 1,000 Watts.) KVA represents the total energy (electricity) provided by the utility to the customer.

The difference between kVA and kW is: in kVA the energy required to energize a magnetic electrical field (i.e., motor windings, transformer windings, etc.) before actual work accomplished is measured. In kW, only the actual demand in Watts is measured and/or averaged. In general, in most electric utilities' rate schedules, the closer the customer's actual kW is to the utilities' kVA, the lower will be the customer's incremental electricity cost.

ITEM #12—Billing Demand

This figure (1,614 kW) represents an actual demand established by the customer in some past billing period-in this sample the 10-01 billing period. This type of demand calculation (actual 1,404 kW vs. billing 1,614 kW) is used by many electricity utilities. The rationale for the utility is that they (the utility) have to maintain and have in reserve the generation capacity the customer needs, whenever the customer needs it.

If this is true, and it is, the utility will take some past billing periods, generally the immediately past 11 billing periods, and bill the customer for the highest actual demand calculated for any of these periods regardless of the current billing period's actual calculated demand. This type of arrangement is referred to as a "trailing ratchet." For each new billing period, the 12th previous billing period is deleted and the current period is included in the billing demand calculation.

The theory in this type of demand billing is that the customer should have an incentive to use their demand on a more uniform basis to reduce the potential billing period's billing demand penalties that occur in any billing period where actual calculated kW is lower than billing kW.
In this sample, the actual calculated kW (1,404 kW—Item #7) compared to the billing kW (1,614 kW—Item #13) results in a penalty of (11614 kW – 1,404 kW) = 210 kW. Using the demand cost of $10/kW shown in Item #17, the penalty is (210 kW × $10) = $2,100. If the customer understands the billing process, this $2,100 should cause them to seriously evaluate their demand characteristics. This $2,100 provides (0) value to the customer. It would be better for the customer to try to determine a demand

reduction strategy and use the $2,100 to reduce demand variability that would have tangible value.

Note: On this sample billing the $2,100 penalty is incrementally identified *nowhere* on the billing recap. The customer has to know what the bill is telling them. If they do not, they may not even be aware of this penalty.

ITEM #13—Contract Demand

Contract demand is just what its name implies a contract, in writing, between the utility and customer establishing some minimum demand that will be billed, or in some instances, a maximum demand that will be allowed over which a penalty will be charged. Provisions like this are required by some utilities so that they are assured of some minimum monthly revenue from the customer to at least partially recover the customer onsite transformation equipment costs.

Are these requirements justified? Probably. But remember, any utility rate schedule provision has to stand the test of the state regulatory rate case procedural process before a utility can actually bill for any rate provision.

In this particular billing, the minimum demand that would be billed would be 1,200 kW, even if the actual kW demand were lower than 1,200 kW. Generally, any provisions like this will have been agreed-to by both parties through a written contract.

ITEM #14—Billing Summary

This area of the billing identifies the cost details that are applicable to the demand (kW) and usage (kWh) characteristics as quantified in Items #1-#13 previously

ITEM #15—Customer Charge

This charge, in this example $210 is what the utility calculates it costs them to provide services to a typical customer in this classification. Included in this charge can be items as follows:

1. Meter reading cost.
2. Meter cost, maintenance, etc.
3. Billing processes.
4. Miscellaneous customer services.

This charge, as all other billing charges, must be approved by the appropriate state regulatory agency before they can be utilized by the utility. Depending upon the utility, customer charges can range from $0 to

$5,000+ per month. Customer charges are generally different for each rate that the utility utilizes and are not negotiable as such. The only reduction in customer charge costs that could generally be experienced would be if a customer were to combine two separate meter points (accounts) into one meter point (account).

ITEM #16—Energy Charge (kWh)

This portion of this billing is the only part that addresses actual electricity energy usage. Out of a billing total of $52,222.26 (Item #27), total energy usage (kWh) is only $24,912.66 ($16,559.64 + $6,891.78 + $1,461.24) or 48% of the total billing amount.

This sample billing has what is called a ratchet energy usage billing provision. This type of usage (kWh) charge is based upon its relationship to the billed demand (kW). In this sample, there are three steps of cost:

<u>1st Step</u>	200 kWh × Billing Demand (1,614 kW)
<u>2nd Step</u>	100 kWh × Billing Demand (1.614 kW)
Remainder	All other kWh

A rate of this type really penalizes a customer with high demand (kW), since a high demand number (kW) increases the number of kilowatt-hours (kWh) in the first step that has the highest per (kWh) incremental cost (in this example is ($.0513/kWh). What the utility is telling the customer through this rate design is that not only will demand (kW) have its own incremental cost, ($10/kW) (see Item #17), but the demand level will also impact the (kWh) incremental cost ($.0513/kWh) in the first step in this example.

Making this type of rate even worse is the fact that the demand (kW) is also ratcheted, and in this example is costing this customer $2,100 (1, 614/kW – 1,404/kW = 210 kW × $10/kW). (See Items #7, #12, #17) This extra 210 kW impacts the usage (kWh) cost also since it places an additional 42,000 kWh (210 kW × 200 = 42,000 kWh) in the first step of the kWh cost ratchet.

If all of this sounds complicated, it is. The thing to remember is that if you are served under a rate like the one in this sample, make certain that you watch both demand (kW) peak levels (Item #7) as well as demand (kW) variability (Item #12) from month-to-month.

ITEM #17—Demand Charge

We have already discussed demand (kW) to some extent with regard to what it is. The demand charge multiplies the billing demand (kW) - in this example 1,614 kW—by the incremental unit cost of demand (kW)—in this example $10.00/kW.

ITEM #18—Fuel Cost Adjustment

This type of charge is typical in electricity billing and represents the utility's (actual) cost of procuring fuel for the electricity generation process. Fuels can include natural gas, petroleum distillates, coal or nuclear materials. The fuel cost is supposed to represent the utility's actual cost of purchasing the generation fuels—generally with no added utility mark up or margin. All fuel costs that are passed on to customers have to be approved by the appropriate regulatory agency.

ITEM #19—Power Factor Adjustment

This adjustment or added cost is based upon the relationship between two units of electricity measurement (kVA) and (kW) (see Item #11). This *adjustment* is actually a penalty in disguise. The ($.7000) figure used here is not indicative of any particular cost calculation other than it is the cost spelled out by the rate schedule.

When electricity billings are analyzed and it is found that costs are being assessed for power factor problems, it will be necessary to investigate the potential for power factor correction procedures as outlined in the power factor section of this manual.

It is important to remember that power factor costs buy you nothing! They are like demand costs but are worse because you can generally eliminate them through corrective measures. If you are paying for power factor problems each billing period, it would be much better to utilize those expenditures to correct the problem and at least end up with something tangible for your money.

ITEMS # 20 & 21—Facility Charge/Facility Lease

When present on a utility billing, these charges generally indicate a nonstandard condition. Usually these types of charges reflect equipment purchases and/or leases by the customer from the utility. In the right situation, these types of arrangements can benefit a customer. The thing that has to be considered by the customer is: do these changes ever end? And generally they do not.

A facility lease, and what is called simply a facility charge, are both illustrated in this example. On an annual basis, these costs are as follows:

1. Facility Charge—$1,147.23/billing period or $13,766.76 annually ($1,147.23 × 12)

2. Facility Lease—$968.17/billing period or $11,618.04 annually ($968. 17 × 12)

These two charges total $25,384.80 annually ($13,766.76 + $11,618.04) on this particular sample billing these charges total $2,115.40, ($1,147.23 + $968.17) or 4% of the total billing amount ($2,115.40 ÷ $52,222.26). While this, as a percentage of total billing cost is relatively small, if it can be eliminated/reduced, it should be.

For both charges, the first thing to do is find a copy of the facility charge/lease written agreement (contract) and determine what the "buy out" value is for the equipment being charged for or leased. If the contract copy cannot be located, contact the utility and ask them for a copy of the particular agreement. For example, let's assume the values for these items are as follows:

1. Facility charge equipment—buyout value $25,000
2. Facility lease equipment—buyout value $20,000

Based upon these example buyout figures, the facility charge equipment could be purchased for less than (22) billing period payments ($25,000 ÷ $1,147.23). The facility lease equipment could be purchased for less than (21) billing period payments ($20,000 ÷ $968.17). If the customer can maintain this equipment or contract for a third-party maintenance agreement, then serious consideration should be given to a "buyout" of the equipment assuming it still has value to the facility.

ITEM # 22—Special Public Utility Commission (PUC) Charge

This charge is imposed by the utility regulatory agency, in this case the public utility commission (PUC). Its purpose is to provide revenue for the regulatory commission for either its normal or some special activity.

Some regulatory commissions receive state financing and some commissions have to fund themselves through individual utility billing charges. Since this charge is listed special, it is likely that it is a special assessment for some particular purpose. Whatever the reasons for this charge, the customer has to pay it as it cannot be negotiated away. The particulars of a charge of this type can be determined by contacting the specific regulatory agency involved.

ITEM # 23—Nuclear Regulatory Cost

This charge, like the special PUC charge (Item #22), is a regulatory commission charge. This particular charge appears to be related to a nuclear (generation reactor) cost of regulation, probably the decommissioning of a particular nuclear reactor site. Whatever its reason, it cannot be negotiated away. The particulars of a charge of this type can be determined by contacting the specific regulatory agency involved.

ITEM # 24—School Tax

This is a local (county, city, etc.) charge for support of the local school system. Even though this charge appears on a utility billing, the actual revenue is given to the school district by the utility. A tax of this type is local in nature and would only be applied to utility billings in a school's jurisdictional area. This type of tax cannot be negotiated away. The particulars of a charge of this type can be determined by contacting the specific county, city, etc. imposing the school tax.

ITEM # 25—Franchise Fee

This fee is actually a local city tax levied upon the utility for the utility's privilege of serving its (the utility's) customers that are within the city's jurisdictional area. The utility then passes this city tax along to the respective customers in this jurisdiction. This tax, like the school tax, is local in nature and cannot be negotiated away. The particulars of a charge of this type can be determined by contacting the specific locality imposing the franchise fee.

ITEM # 26—State Sales Tax

This is a statewide tax that is not local in nature. It applies equally to all utility customers unless there are specific sales tax exemptions that can apply to a specific customer. To determine if there are any existing exemptions and their applicability, contact the state department of taxation or revenue. Do not contact the utility since they cannot help in this matter. If there are state sales tax exemptions and they are applicable, the state will issue an exemption number that will need to be submitted to the utility for the purpose of having the tax reduced/removed from the monthly billing.

An important feature of most state sates tax exemption provisions is that a customer can get back state sales tax funds paid in the past if the customer legally was not required to pay the tax. The refund period, based upon specific state statute of limitation provisions, generally are from three to five years in the past.

If it is apparent that unnecessary state sales tax has been paid, it is the customer's responsibility to contact the state and take the appropriate action to have the tax reduced/eliminated. It is not the state's responsibility and certainly not the utility's responsibility to take this action. The state assumes that all utility customers owe the state sales tax. It is the customer's responsibility to prove otherwise.

When exemptions exist, they generally apply to the following types of customers:

1. Not-for-profit entities
2. Certain food service/manufacturing facilities
3. Several manufacturing facilities
4. City, county, state agencies

The particulars of a charge of this type and the potential for exemption can be determined by contacting the specific state imposing the sales tax.

ITEM # 27—Total Current Bill

This is a totalizing of the charges listed in Items #15-#26. In this example, there are no past due charges so the amount in this item represents the actual total billing period charges for this period—$52,222.26.

ITEM # 28—Previous Amount Due

This is the billing amount of the previous billing period in this example—$49,176.83. In the next billing period, this previous amount due figure will be $52,222.26, or this billing period's total current bill amount. (See Item #27)

ITEM # 29—Payment Received

In this example, the customer paid the previous billing amount due ($49,176.83), so the payment received matched the previous amount due. If the customer had not paid the previous amount due, any discrepancies would appear here.

ITEM # 30—Total Amount Due

This figure is a combination of the total current billing plus the previous amount due, minus the payment received. In this example, the figure is \$52,222.26 – (\$52,222.26 + \$49,176.83 – \$49,176.83). Any non-payments of previous amounts due win be added to the total current bill and shown here.

While the explanation of this one page example billing (Figure 3.1) may seem very involved and perhaps overly complex, it is critical that the customer thoroughly understands what their billing is telling them. If a customer does not understand their utility billing, they have little hope of reducing their utility costs.

AREAS TO INVESTIGATE FOR COST REDUCTION OPPORTUNITIES ON FIGURE 3-1 BILLING

1. **9→** RATE GSP-3

Is this the most cost effective rate available based upon current usage characteristics?

2. **11→** POWER FACTOR 71%

19→ POWER FACTOR ADJUSTMENT
(\$.7000 × 1,614 kW) = \$1,129.80

Poor power factor (efficiency of usage) is costing \$1,129.80 or \$13,557.60 yearly. Investigate installation of power factor improvement capacitors.

3. **16→** ENERGY CHARGE

1st 200 × 1,614 kW = 322,800 kWh × \$.0513/kWh	\$16,559.64	
2nd 100 × 1,614 kW = 161,400 kWh × \$.0427/kWh	\$6,891.78	\$24,912.66
Remainder = 73,800 kWh × \$.0198/kWh	\$1,461.24	

Energy charges are based upon the billing demand level (1,614 kW) or a total cost of \$24,912.66. If the billing and actual peak demand levels would have been the same (1,404 kW), the total energy charges would have been as follows:

Energy Charge

1st 200 × 1,404 kW = 280,800 kWh × .0513/kWh	$14,405.04	
2nd 100 × 1,404 kW = 140,400 kWh × $.0427/kWh	$5,995.08	$23,108.76
Remainder = 136,800 kWh × $.0198/kWh	$2,708.64	

The difference between what was paid for energy ($24,912.66) and what could have been paid ($23, 108.76) results in a difference of ($1,803.90) for this month; or, ($21,646.80) yearly. The cause of this difference is the large variability in peak demand.

4. **17→** <u>DEMAND CHARGE</u>
(1,614 kW × $10.00/kW) = $16,140.00

The actual peak demand this month was 1,404 kW. (Item #7 on Billing Detail) Due to a ratchet clause the demand paid for was 1,614 kW - a difference of 210 kW. (210 kW × $10.00/kW) = $2,100; or, ($25,200.00) yearly. Investigate reasons for large variations in peak demand levels.

5. **20→** <u>FACILITY CHARGE</u> $1,147.23

6. **21→** <u>FACILITY LEASE</u> $968.17

These charge/lease costs relate to some type of special facility equipment required by the customer but not provided by the utility as a standard *no cost* item. This cost ($2,115.40) for this month; or, ($25,384.80) yearly should be investigated as follows:

a) Are the facility charge and facility lease necessary?

b) If the facility charge and lease are required, could the facilities be purchased to eliminate the monthly charges/lease fees?

SYNOPSIS—ANALYZING THE ELECTRIC UTILITY BILL

A thorough knowledge of the utility billing detail is needed to be able to understand the current billing process, as well as determining the potential for alternative billing opportunities.

Total Cost Reduction Opportunity Items on Figure 3-1 Billing

	Item	*Item #*	*Amount*
1.	Rate	(9)	Unknown
2.	Power Factor	(11/19)	$1,129.80
3.	Energy Charges	(16)	$1,803.90
4.	Demand Charge	(17)	$2,100.00
5.	Facility Charge & Facility Lease	(20/21)	$2,115.40
		Total for this Month	$7,149.10
		YEARLY TOTAL	$85,789.20

These opportunities represent 15.6% of the total energy cost: *$45,791.26 (*billing items #15, 16, 17, 18, 19, 20, 21).

Chapter 4

Electricity Billing Components

ELECTRICITY BILLING DETAILS

To be able to reduce electricity costs, the billing components that impact cost must be understood. There are five standard items that make up the composite cost total.

Each of these items is different and their impact on the total electricity cost depends upon the specific serving utility. Only by understanding each of these items will the potential for maximizing cost reductions be realized. The five billing components are:

1. Rate
2. Demand
3. Power factor
4. Voltage level
5. Usage

I. RATE

- Rate suitability is the customer's responsibility
- Savings potential: up to 30%

Cost Components of Electricity—Rate

Most cost-effective rate is the customer's responsibility. There may be more than one rate that can be utilized for a given demand and usage combination, but only one rate will be the most cost effective.

Steps in Evaluating Rate Appropriateness

1. Obtain complete utility rate schedule (all rates available).

2. Determine rates that could be applicable to your usage characteristics:
    - Demand
    - Usage levels
    - Hours
    - Days worked
    - Future facility changes/revisions

3. If an alternative rate is found that *appears* to be viable, have the utility service representative calculate the cost for your usage characteristics for the past 12 months. If the cost is less than what was paid on the current rate, consider changing to the alternative rate. Before changing rates, make certain that the next 12 months usage characteristics will be similar to the previous 12-month sample period.

Figure 4-1 Explanation—(Rate)
Rate: GS—General Service
1. Are other rates available?
2. Are non-standard rates available?
3. Are experimental rates available?

Rate Overview
1. Look for alternative rates.
2. Evaluate potentials for cost reductions.
3. Consider future usage characteristics.
4. Use alternative rate, if appropriate.

II. DEMAND

- Demand is *capacity reservation charge.*
- Savings potential: 5% to 20%

Cost Components of Electricity—Demand

Demand, as it applies to an electrical system is defined as—*the rate at which electric energy is delivered to or by* a *system, part of* a *system, or* a *piece of equipment expressed in kilowatts (kW), kilovolt-amperes (kVA), or other suitable unit* at a *given instant or average over* a *designated period of time.*

The primary source of demand is the power-consuming equipment of a customer. Peak or maximum demand charges are applied to the maximum

RATE

UTILITY, INC.	STATEMENT OF ELECTRIC SERVICE	ACCOUNT NUMBER
For Inquires 24 Hours per Day	COMPANY NAME:	DATE DUE:
For Payment Locations Call:	BILLING ADDRESS	TOTAL AMOUNT DUE: $5,920.82
Website:		NEXT READ DATE ON OR ABOUT:
To Report a Power Outage:	SERVICE ADDRESS:	DEPOSIT AMOUNT ON ACCOUNT:
— METER READINGS —	GS — GENERAL SERVICE	*SECONDARY VOLTAGE (220 / 440-3 PHASE)*

METER NUMBER		0672132
KWH PRESENT	(Actual)	044335
KWH PREVIOUS	(Actual)	043127
DIFFERENCE		1208
CONSTANT		40
TOTAL KWH		48320
KW PRESENT	(Actual)	4.175
CONSTANT		40
TOTAL KW		167
KVA PRESENT	(Actual)	6.95
CONSTANT		40
TOTAL KVA		278

– ENERGY USE–	
DAILY AVERAGE USE	1611 KWH / DAY
USE ONE YEAR AGO	1439 KWH / DAY

BILLING PERIOD:	JAN 01 TO JAN 31		30 DAYS
CUSTOMER CHARGE		21.64	
ENERGY CHARGE INCLUDING FUEL COST ADJUSTMENT	48320 KWH @ 5.1263¢	2,477.03	
DEMAND CHARGE	250 KW @ 9.5299¢	2,382.47	
TOTAL ELECTRIC COST			$4,881.14
GROSS RECEIPTS TAX	@ 2.5%		122.03
MUNICIPAL FRANCHISE FEE	@ 6.5%		317.27
MUNICIPAL UTILITY TAX	@ 3.6%		175.72
SALES TAX	@ 8.7%		424.66
TOTAL CURRENT BILL			$5,920.82
TOTAL DUE THIS STATEMENT			$5,920.82

Payment of your bill prior to the above due date will avoid a late payment charge of 1.5%

OUR 24-HOUR OUTAGE NUMBER is the quickest way to report power outages – Just Call:

Figure 4-1. Rate—Sample Billing

demand for energy required by a system in a given period of time. Utilities normally charge a monthly fee based upon peak demand. This is peak demand generally expressed as kilowatts (kW) kilovolt-amperes (kVA) and averaged in a given period of time—either a 15- or 30-minute interval.

This peak or maximum demand charge can vary from less than $2.00 to $18.00 or more per kilowatt per month. Reduction of demand peaks can result in sizable savings. Many times a revision in how and/or when equipment is turned on or off can be all that is needed to reduce monthly demand charges. In other instances, a computer controlled energy management system can be installed which will sense and adjust changing energy requirements to reduce peak demands.

How Demand Occurs

Demand (kVA/kW) occurs as a result of multiple electrical consuming items operating at the same time. A utility customer is charged for the highest (peak) kVA/kW in a 15/30 minute averaged time period during a billing cycle. Other factors can impact the actual kVA/kW charges a customer may actually pay; i.e., ratchet factors, minimum contract billing provisions, etc.

Following is an example of how demand occurs. Demand is present anytime electricity is being consumed. This example shows how individual demand levels affect kVA/kW peak totals.

Figure 4-2. Demand Components (Peak Period)

	Item	*kVA/kW Averaged Over 15/30 Minute Period*
1.	Motor	97
2.	Motor	63
3.	Motor	51
4.	Motor	25
5.	Motor	20
6.	Lighting	114
7.	Lighting	97
8.	Lighting	90
9.	HVAC	170
10.	HVAC	92
11.	Office	85
12.	Miscellaneous	87
	Total	991 kVA/kW

On the example shown, the customer would pay for 991 kVA/kW, assuming no other rate provisions applied.

Explanation of Figure 4-3, Demand Profile-Time-of-Use/Time-of-Day

This actual demand profile is the same as the one shown in Figure

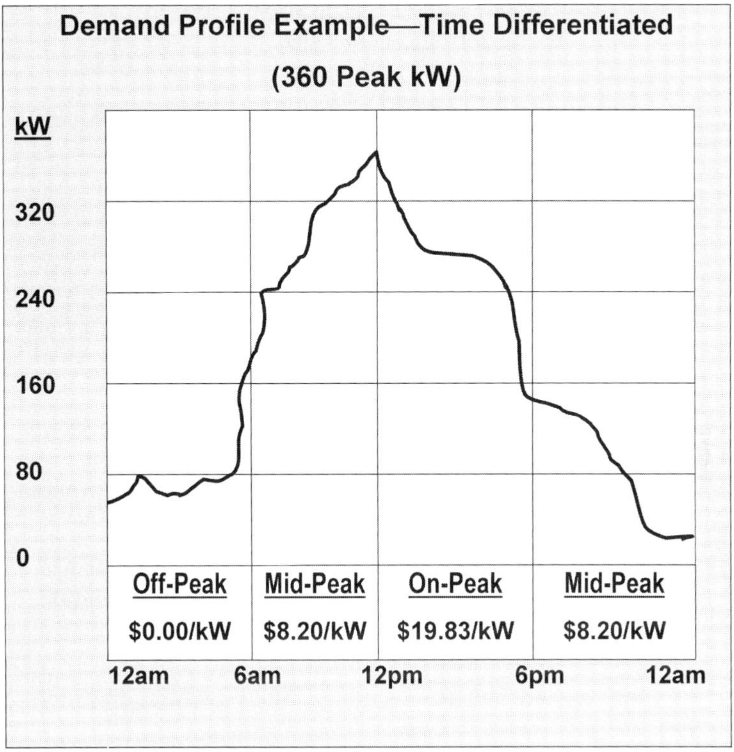

Figure 4-3. Analyzing Time Differentiated Demand Profiles vs. Non-time-of-use/Time-of-day Demand Profiles

4-4. The difference between Figure 4-3 and Figure 4-4 is in the method utilized by the utility to charge for the peak demand units. In this figure, there are three different demand charges applied to actual demand depending upon which time period it occurs in as follows:

1. Off-peak (12 a.m.-6 a.m.)—$0.00/kW
2. Mid-peak (6 a.m.-12 p.m.) and (6 p.m.-12 a.m.)—$8.20/kW
3. On-peak (12 p.m.-6 p.m.)—$19.83/kW

In Figure 4-3, actual demand (kW) unit costs range from ($0.00 off-peak to $19.83 on-peak). In a utility billing scenario like the one shown in this figure, there can be great savings if the demand peak can be shifted. In Figure 4-4, the demand unit cost is $19.83/kW regardless of when the demand peak occurred. Also in Figure 4-4, demand unit cost varies greatly from ($0.00 to $19.83/kW) depending upon when it occurs.

On both figures, the demand peaks at 12 p.m. noon. In Figure 4-3, the demand peak at 12 p.m. noon is in the most expensive period (on-peak—$19.83/kW). In Figure 4-4, demand cost is no greater or less than at any other time.

If through scheduling changes, equipment revisions, etc., the demand peak could be moved to sometime prior to 12:00 p.m. (11:55 a.m., 11:45 a.m., 11:30 a.m., etc.), the demand unit cost would be reduced to $8.20/kW. In Figure 4-3, the total utility demand charge could fall from $7,138 (360 kW × $19.83/kW) to $2,952 (360 kW × $8.20/kW) or a potential savings of $4,186 ($7,138 – $2,952).

Remember, demand (kW) charges purchase only generation capacity reservation-not any actual electricity usage (kWh). In Figure 4-3, a customer must determine whether it would be possible to move (shift) their peak demand to some other less costly billing period based upon potential savings that could result—in this case $4,186.

Reduction of peak demand levels, while not included in Figure 4-3, could result in even greater savings than those shown. Demand, while a component of electricity purchase for practically all customer classes other than residential and perhaps small commercial categories, should be evaluated frequently for potential shifting and/or reduction potentials.

Explanation of Figure 4-4
Demand Profiles—Non-time Differentiated

This demand profile and the one shown in Figure 4-3 are identical. The difference is in how the demand charges are assessed. In Figure 4-4, the actual demand charges are what are called non-time differentiated.

What this means is that the utility's charge per unit of peak demand is the same or uniform regardless of when the actual demand peak occurs. If the actual demand peak occurs in the morning, afternoon, evening, weekend, day or holiday, the utility's charge per demand unit (kW) will be the same—$19.83/kW.

When demand units are billed in this manner, there is not an opportunity to shift the actual demand to reduce costs. With this type of billing, the only way to reduce demand charges is to reduce the actual demand units being registered on the customer meter.

Steps That Can Assist in Determining and Correcting Peak Demands

1. **Determine** current peak demand and the monthly charges related to

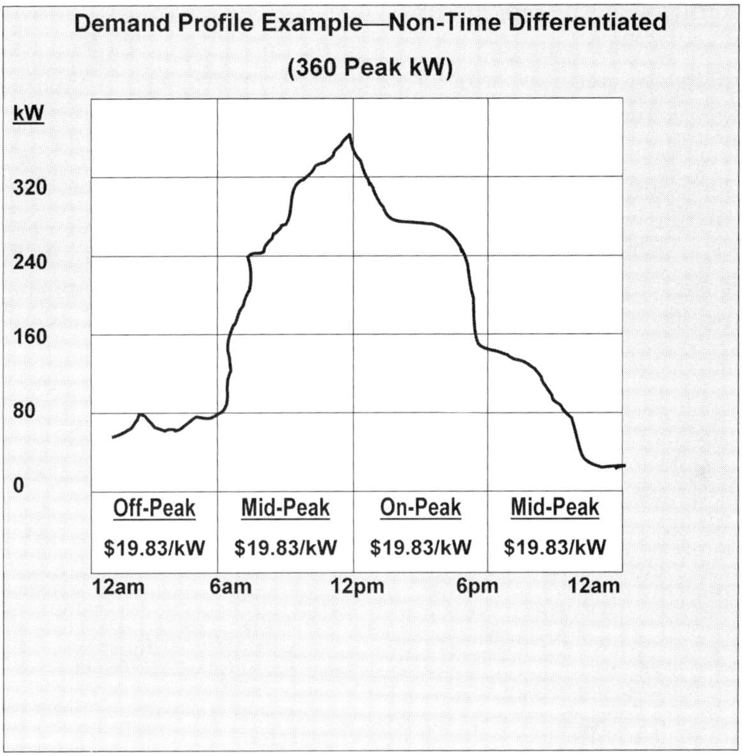

Figure 4-4. Analyzing Time Differentiated Demand Profiles vs. Non-time-of-use/Time-of-day Demand Profiles

it. (This information can generally be obtained from the monthly utility bill.) Make certain that power factor or ratchet provisions are understood and taken into consideration when analyzing demand data.

2. **Contact** the utility and request that a strip chart recorder be installed in the system for at least a one-month period. The purpose of this recorder is to document, in strip chart form by day and time-of-day, the variations in the electrical demand of your operation. When the recorder chart strip is received, it will look very similar to an electrocardiogram in that it will show the peaks and valleys caused by changing demands.

   When the chart strip is analyzed, try to determine if there are any repetitive peak patterns, from hour-to-hour, day-to-day, or week-to-week. If there are repetitive patterns, determine what is happening at those times that cause the peaks to occur; e.g., start of a work shift, employee break or lunch times, equipment testing, etc.

3. **Develop** the corrective actions that need to be made after the information is received regarding the peak periods. Try to reduce, perhaps the top 25% or 30% of peak demand periods. This is not only the less costly approach but in cases can be accomplished by either procedural change in equipment usage or personnel schedules. Work with the utility company since they can provide much technical insight into how to lower specific peak demand periods.

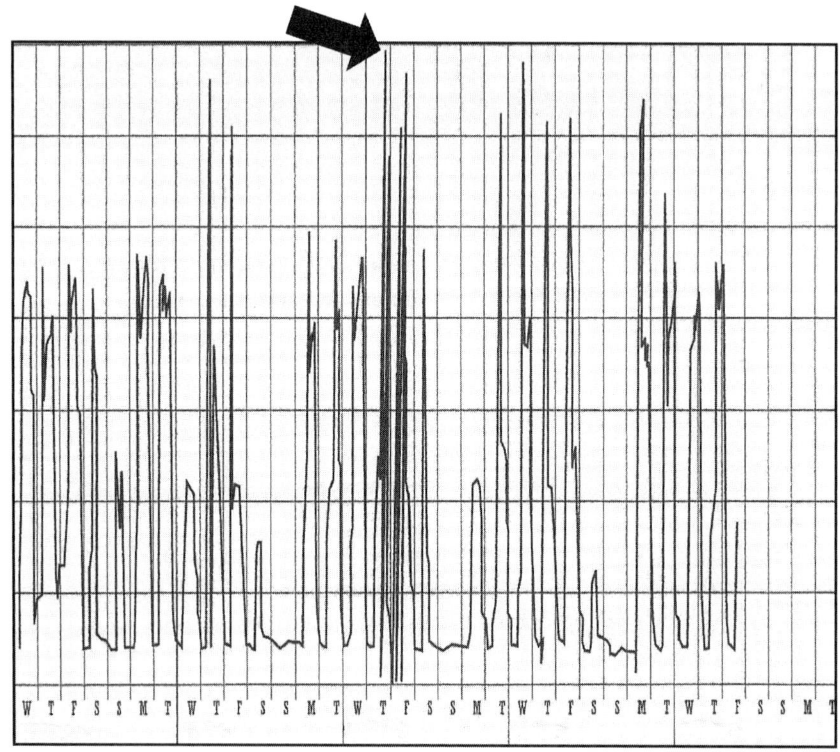

Figure 4-5. Analog Plotting of Digital Data

Explanation of Figure 4-5—Analog Plotting of Digital Data

This type of chart is not really originated by or through actual measurement of demand levels but is calculated by processing the digital data actually measured through conversion of digital to analog data. The reason for utilizing an analog chart is the amount of information that can be shown on one chart.

Item #1—Days/Weeks

This chart is designed for a 4-week period, which is generally sufficient for a utility billing period. Every day of every week is shown on this chart.

Item #2—Demand Levels

This column simply shows the demand threshold numbers that apply to the chart.

Item #3—Peak Demand

The peak demand on this chart peaked at 3,490 kW. Subject to applicable utility rate provisions, this single peak demand level could determine the total demand charge for the billing period.

The information that this charting format shows is the variability between daily demand peaks. This information, if properly understood and utilized, can lead to potential demand peak reduction strategies. For example in this chart, the following information can be determined:

Question: What does this chart show that would help to determine the minimum demand peak level that might be possible to obtain?

Answer: Probably the ultimate goal would be to reduce demand levels to (0) but from a practical standpoint, this would never happen unless the business discontinued operation—not a very viable option to reduce demand charges.

Question: What is this chart telling us with relation to how much peak demand could be reduced "practically?"

Answer: If we study the data on the chart, it appears that for this particular billing period, any reduction below 3,000 kW would be difficult to implement if for no other reason than the multiple times peak demands occur in the 2,500/3,000 kW interval. If we analyze the peak demands over 3,000 kW, we find that there are relatively few (11) occurrences above 3,000.

The occurrences above 3,000 kW calculated as a percent of total occurrences over the period is (30 days × 96 peak periods/day = 2,880) total occurrences—11 occurrences over 3,000 kW + 2,880 total occurrences = .0037 of all occurrences were over 3,000 kW.

If we assume that this demand profile is representative of what occurs every month, and if we use a $10/kW charge, then the 490 kW over 3,000 kW cost is 490 kW × $10 = $4,900. If we further project this cost on an annual basis, we have $4,900 × 12 months = $58,800.

The question an electricity customer should ask themselves in a situation like this is: would it be better to take the potential $58,800 that would have been spent on demand over 3,000 kW on a yearly basis and develop a strategy to eliminate demand peaks over 3,000 kW? When demand is purchased, it accrues no value and it has to be paid for each month that it occurs.

EXAMPLE #1

ON-PEAK HOURS: 12 noon-8 p.m., Monday-Friday.
INTERMEDIATE HOURS: 8 a.m.-12 noon and 8 p.m.-12 midnight, Monday-Friday.

OFF-PEAK: All other hours including the holidays of New Year's Day, Rev. Martin Luther King's birthday, President's Day, Memorial Day, Independence Day, Labor Day, Columbus Day, Veteran's Day, Thanksgiving Day, and Christmas Day as designated by the federal government, shall be considered off-peak.

EXAMPLE #2

At customer's option, the ON-PEAK hours are 7 a.m.-3 p.m., 8 a.m.-4 p.m., or 9 a.m.-5 p.m., Monday-Friday, inclusive, New Year's Day, Memorial Day, Independence Day, Labor Day, Thanksgiving Day, and Christmas Day.

EXAMPLE #3

ON-PEAK HOURS: 8 a.m.-8 p.m., Monday-Friday. All other hours will be considered off-peak. Utility reserves the right to change the on-peak hours from time-to-time. The off-peak hours will not be less than 12 hours daily.

NOTE: This rate provision could be the most costly one on this sheet from the customer's point-of-view since the utility has the right to shift on-peak hours anytime period of the day so long as off-peak hours are not less than 12 hours daily.

How does a customer shift their hours of operation to match a revised on-peak period?

Figure 4-6. Examples of Rate Schedule Provisions (On-peak, Intermediate, and Off-peak Demand Billing Hours)

EXAMPLE #1—

<u>*Billing Demand*</u>: The maximum 30-minute measured demand in the month.

EXAMPLE #2—

<u>*Billing Demand:*</u> The maximum 15-minute measured demand in the month, but not less that 10 kW.

EXAMPLE #3—

<u>*Billing Demand:*</u> The greatest of the following:

A. The maximum 15-minute measured kW demand in the month.
B. 80% of the maximum 15-minute measured kVA demand in the month (power factor clause).
C. The maximum demand as so determined above during the preceding 11 months (ratchet clause).
D. 50 kW.

EXAMPLE #4—

<u>*Billing Demand:*</u> The maximum 15-minute measured demand during the on-peak hours of 8 a.m.-8 p.m., Monday-Friday. The off-peak demand shall be the maximum demand created during the remaining hours (time-of-use/ time-of-day).

EXAMPLE #5—

<u>*Demand Charge*</u>: $90.00 for first 20 kW demand or less; and, $3.10 per kW for all additional kW demand.

EXAMPLE #6—

<u>*Demand Charge*</u>: $5.00 per kW demand.

EXAMPLE #7—

<u>*Demand Charge*</u>: Base charge plus adder based upon voltage levels:

$12.73 per kW on peak billing demand, plus.
$3.75 per kW maximum demand (less than 24 kV) primary service
$2.75 per kW maximum demand (24 to 41.6 kV) subtransmission voltage level
$1.50 per kW maximum demand (120 kV and above) transmission voltage level

Figure 4-7. Examples of Rate Schedule Provisions Regarding Demand Billing Procedures

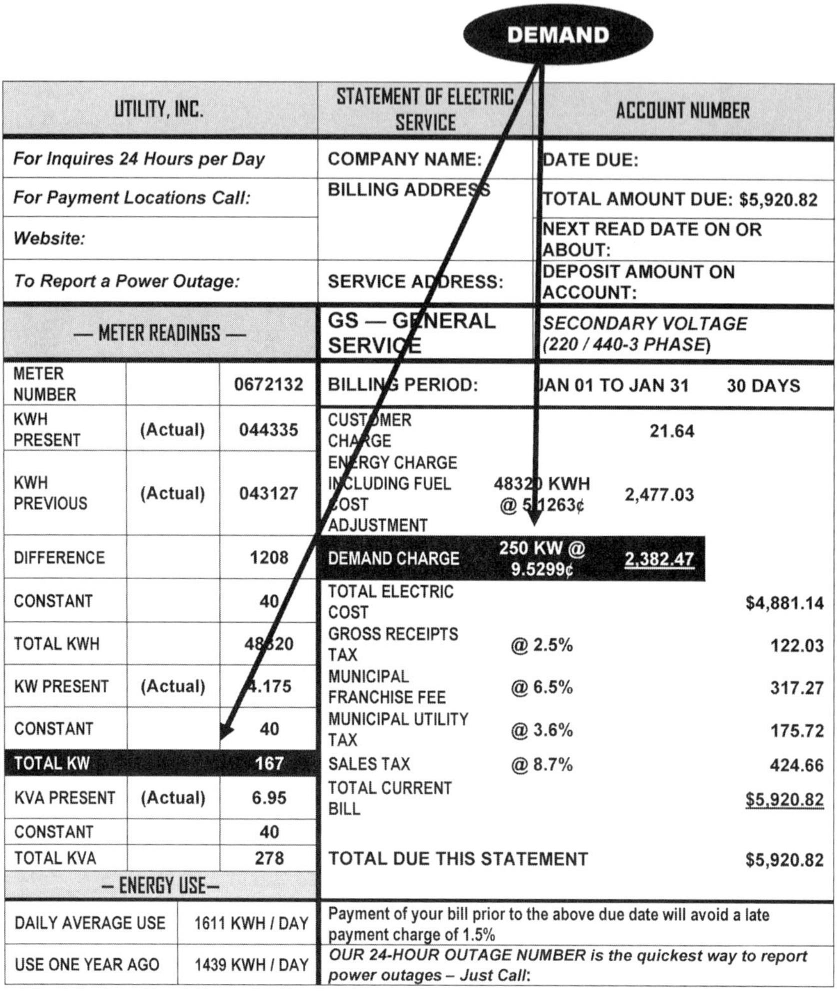

UTILITY, INC.	STATEMENT OF ELECTRIC SERVICE	ACCOUNT NUMBER
For Inquires 24 Hours per Day	COMPANY NAME:	DATE DUE:
For Payment Locations Call:	BILLING ADDRESS	TOTAL AMOUNT DUE: $5,920.82
Website:		NEXT READ DATE ON OR ABOUT:
To Report a Power Outage:	SERVICE ADDRESS:	DEPOSIT AMOUNT ON ACCOUNT:

— METER READINGS —		
METER NUMBER		0672132
KWH PRESENT	(Actual)	044335
KWH PREVIOUS	(Actual)	043127
DIFFERENCE		1208
CONSTANT		40
TOTAL KWH		48320
KW PRESENT	(Actual)	4.175
CONSTANT		40
TOTAL KW		167
KVA PRESENT	(Actual)	6.95
CONSTANT		40
TOTAL KVA		278

GS — GENERAL SERVICE	*SECONDARY VOLTAGE (220 / 440-3 PHASE)*		
BILLING PERIOD:	JAN 01 TO JAN 31	30 DAYS	
CUSTOMER CHARGE		21.64	
ENERGY CHARGE INCLUDING FUEL COST ADJUSTMENT	48320 KWH @ 5.1263¢	2,477.03	
DEMAND CHARGE	250 KW @ 9.5299¢	2,382.47	
TOTAL ELECTRIC COST			$4,881.14
GROSS RECEIPTS TAX	@ 2.5%		122.03
MUNICIPAL FRANCHISE FEE	@ 6.5%		317.27
MUNICIPAL UTILITY TAX	@ 3.6%		175.72
SALES TAX	@ 8.7%		424.66
TOTAL CURRENT BILL			$5,920.82
TOTAL DUE THIS STATEMENT			$5,920.82

— ENERGY USE—	
DAILY AVERAGE USE	1611 KWH / DAY
USE ONE YEAR AGO	1439 KWH / DAY

Payment of your bill prior to the above due date will avoid a late payment charge of 1.5%

OUR 24-HOUR OUTAGE NUMBER is the quickest way to report power outages – Just Call:

Figure 4-8. Demand—Sample Billing

Figure 4-8 Explanation—(Demand)

Demand Cost

250 kW Billed
167 kW Actual
83 kW Penalty

- 83 kW × $9.5299 per kW = $790.98 PENALTY
- Why was 250 kW billed, but actual was 167 kW? Poor power factor.

(See power factor Figure 4-10.)

- OR 16% of TOTAL COST

Demand Overview

1. Utilize strip chart data
2. Analyze strip chart information
3. Take corrective action

III. POWER FACTOR

- Power factor is efficiency of usage (kVA/kW)
- Savings potential: 5% to 15%

POWER FACTOR—
HOW IT IMPACTS UTILITY GENERATION LEVELS

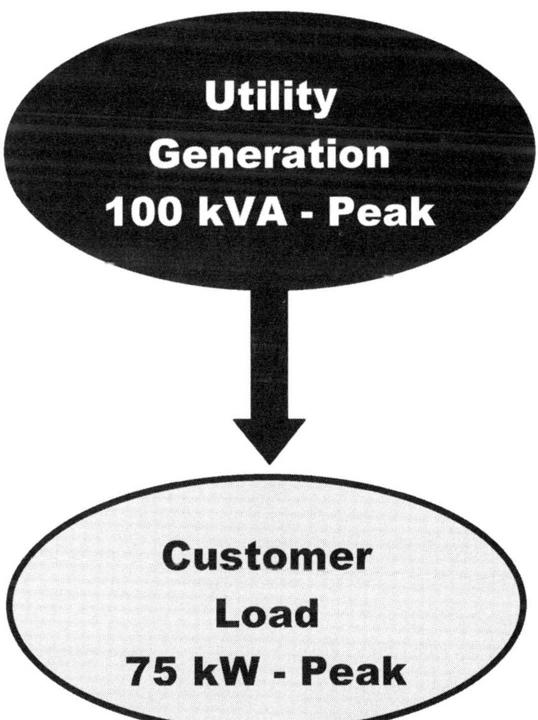

If this utility bills on (kW), they are receiving revenue only on 75% (75 kW) of the actual reservation capacity (100 kVA) that they provide for this customer.

KILOVOLT-AMPERES (kVA) vs. KILO WATTS (kW)

What Is Power Factor and What to Do to Correct It

Power factor, technically stated is the ratio of real power—kilowatts (kW), to apparent power—kilovolt-amperes (kVA) for any given load and time. Generally, it is expressed as a percentage ratio. Simply stated, it is the efficiency at which electricity is consumed.

Since utilities provide their electrical energy in units of kilovolt-amperes (kVA), and a customer's usage is generally measured in kilowatts (kW), the power factor or relative efficiency of the customer's usage may be calculated as in Figure 4-9.

For a Given Billing Period		
1.	Utility provided peak kVA	863 kVA
2.	Customer utilized peak demand in kW	647 kW

The power factor in this example would be
647 kW ÷ 863 kVA = 75%

Figure 4-9. Power Factor Calculation

Since the utility had to provide 25% (100%-75%) more kVA than was represented in the customer's kW peak demand (863 kVA vs. 647 kW), there would probably be a power factor correction penalty that would be detailed in the applicable rate schedule.

Most utilities measure power factor since the inefficient use of electricity requires more power reserves than would be needed if the power were used efficiently. Also, most utility rate schedules provide for the institution of penalty charges if power factor percentages are under minimum figure—normally 85%.

If the power factor ratio falls below the minimum penalty point, it can normally be raised to a non-penalty level by the installation of capacitors that will store the electrical power required during inefficient power

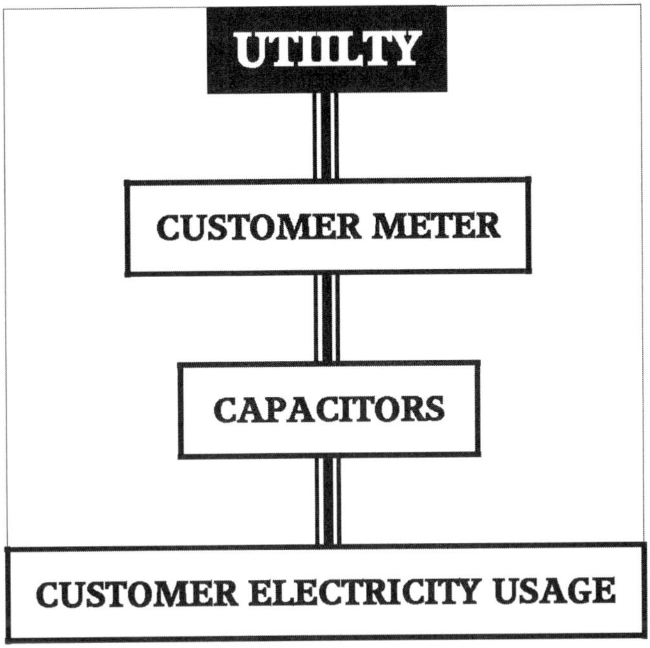

Figure 4-10.
Steps Utilizing a Capacitor Bank to Reduce/Eliminate Power Factor Cost

usage periods.

These capacitors can be in the form of capacitor banks or groups installed at the main electrical distribution point or they can be installed individually at specific areas of inefficient electrical usage. Power factor correction can be a source of considerable savings and generally is rather straightforward in being corrected.

Investigative Steps in Correcting Power Factor

1. Determine if the utility imposes power factor surcharges.
   - This can generally be determined from the monthly billing.

2. If low power factor surcharges are imposed, determine how much these charges are.
   - Discussion with the Utility Service Representatives can help to determine what these charges are if they are not apparent from the monthly utility billing.

3. In conjunction with the utility, determine the amount of capacitance correction that is required for the system.
   - Generally, the utility will install the required capacitors for you if they can be installed at the main electrical distribution point. Naturally, there is a charge for this, but it may be best to have the utility do the installation since they are most familiar with the system. Also, capacitors can be purchased or leased from third party suppliers.

4. Calculate the payback by comparing the monthly low power factor surcharge with the cost of installing the required capacitors.
   - In many cases, the payback will be less than one year.

Figure 4-11 illustrates the steps to utilize individual power factor correction capacitors at individual points of low or poor power factor throughout the customer's electrical distribution system to improve actual system efficiency.

This type of power factor correction addresses only utility-imposed power factor penalty charges, and electrical inefficiencies continue to exist within the customer's electrical distribution system These inefficiencies are manifested through reduced electrical distribution capacity within the customer's system.

Poor power factor on an electrical distribution system is similar to the condition that arises when the arteries and veins of a person become clogged with plaque that results in reduced blood flow. The result in an electrical system is reduced electrical carrying or distribution capacity.

One indication of insufficient electrical carrying capacity in an electricity distribution system is the presence of excessive heat with the system. The method of power factor correction to reduce or eliminate utility surcharges shown in this example represents the procedure most frequently utilized due to the ease and cost of the installation.

Keep in mind that this method only masks the real problem; it does not address the cause of the problem. Power factor penalties are generally easy to determine since they appear on the monthly utility billing so payback periods can be quickly calculated.

This type of correction is the method to utilize to actually address electrical inefficiencies that result in poor power factor problems. Unfortunately, this method is not often used when analyzing existing electrical distribution problems because of the cost and time that might be re-

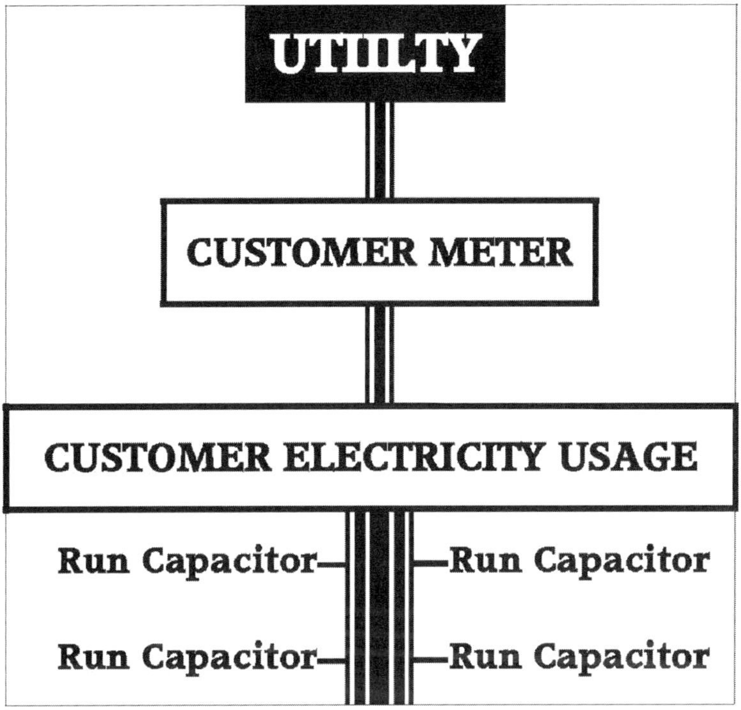

Figure 4-11.
Steps Utilizing Individual Power Factor Correction Capacitors

quired to find all areas of poor power factor. The correct methods to keep an electrical distribution system efficient are as follows:

1. **Design** new electrical systems to some predetermined power factor threshold—85%-90% as an example.

2. **Correct** potential poor power factor areas as changes occur on an existing electrical system.

3. **Survey** the existing electrical system to determine whether any of the following conditions exist so that measures can be undertaken to correct problems.
   A. Intermittent operation of electric motors—periods of electric motor idling time contributes to poor power factor.

B. Oversized electric motors. Example-using a 50-hp motor would be sufficient.
C. Fluorescent lighting using transformer-type ballasts—correct through solid state/energy efficient design ballasts. Note: Many times utility rebate programs assist with the cost related to changing to energy efficient fluorescent lighting.

These items certainly are not all inclusive of the things that contribute to poor power factor, but represent some of the more common problem areas. Probably a good place to start in analyzing any electrical system would be to contact the electric utility Service Representative to determine what assistance the utility could provide in determining low power factor problems. Many utilities have in-house personnel that can assist a customer through an onsite evaluation at no cost. Other times the utility can suggest reputable outside sources that can provide an evaluation of specific needs at reasonable costs.

Capacitor kVAR and Load Capacity

How many capacitors are needed to improve power factor from its present level to any desired value?

Using the power factor improvement table, Figure 4-12:

1. Locate the original power factor on the vertical scale.
2. Trace horizontally across the table to directly below the desired power factor point.
3. Multiply the number found by the load in kilowatts.
4. The result is directly in capacitor kVAR.

Example

- 400 kW at 75% power factor
- Desired improvement is to 95% power factor

1. Locate vertically at 75%—current power factor in % level
2. Locate horizontally at 95%—desired power factor in %
3. Intersect the vertical (75%) and horizontal (95%) lines—.553
4. Multiply .553 × 400 kW = 221 kVAR of capacitors needed

CURRENT POWER FACTOR IN (%) — Multiplying constants	DESIRED POWER FACTOR IN (%) 80%	85%	90%	95%
50%	0.982	1.112	1.248	1.403
55%	.769	.899	1.035	1.190
60%	.584	.714	.850	1.005
65%	.419	.549	.685	.840
70%	.270	.400	.536	.691
75%	.132	.262	.398	.553
80%	.000	.130	.266	.421
85%	.000	.000	.136	.291
90%	.000	.000	.000	.155
95%	.000	.000	.000	.000

Figure 4-12. Power Factor Capacitor Improvement Table

Explanation of Figure 4-13—Actual Example of Power Factor Cost

This example is a representation taken from an actual utility billing. All of the information and billing format are exactly as was illustrated on the original billing. The utility name and customer specific data are the only things that have been deleted.

Item #1—Actual Demand

This item shows the actual peak demand (1,344 kVA) that the customer meter registered during the billing period (Dec 06 to Jan 03). **Note**: the unit of measure is not kW (kilowatts) but kVA (kilovolt-amperes).

This difference is important since it is expressed in (kVA) which represents the actual demand units the utility registers on their generation equipment which may be different from what the customer's actual peak demand in (kW). The kVA numbers incorporate the reactive demand of electrical equipment that results from inefficiencies of the equipment during its operation. If a utility incorporates power factor calculations in its billing process, kVA will always be the baseline against which kW will be measured.

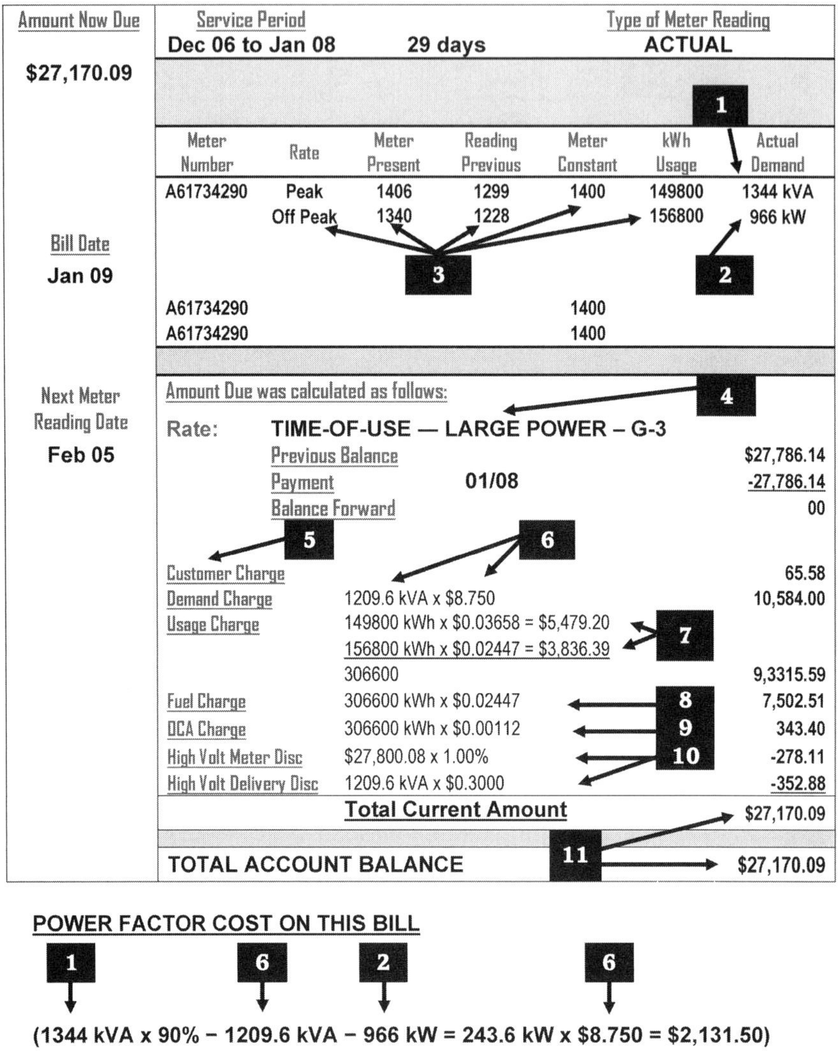

Amount Now Due
$27,170.09

Service Period
Dec 06 to Jan 08 29 days

Type of Meter Reading
ACTUAL

1

Meter Number	Rate	Meter Present	Reading Previous	Meter Constant	kWh Usage	Actual Demand
A61734290	Peak	1406	1299	1400	149800	1344 kVA
	Off Peak	1340	1228		156800	966 kW
A61734290				1400		
A61734290				1400		

3 2

Bill Date
Jan 09

Next Meter Reading Date
Feb 05

Amount Due was calculated as follows: 4

Rate: TIME-OF-USE — LARGE POWER – G-3

Previous Balance		$27,786.14
Payment	01/08	-27,786.14
Balance Forward		00

5 6

Customer Charge			65.58
Demand Charge	1209.6 kVA x $8.750		10,584.00
Usage Charge	149800 kWh x $0.03658 = $5,479.20	7	
	156800 kWh x $0.02447 = $3,836.39		
	306600		9,3315.59
Fuel Charge	306600 kWh x $0.02447	8	7,502.51
DCA Charge	306600 kWh x $0.00112	9	343.40
High Volt Meter Disc	$27,800.08 x 1.00%	10	-278.11
High Volt Delivery Disc	1209.6 kVA x $0.3000		-352.88
	Total Current Amount		$27,170.09

TOTAL ACCOUNT BALANCE 11 $27,170.09

POWER FACTOR COST ON THIS BILL

1 6 2 6

(1344 kVA x 90% – 1209.6 kVA – 966 kW = 243.6 kW x $8.750 = $2,131.50)

Figure 4-13. Actual Example of Power Factor Cost

Notice on this billing, information is absent concerning power factor or if it is even utilized in total cost figure calculations. The only way a customer can calculate a billing of this type is to know what is being shown. Also, notice on this billing, there is no data relating to how the kVA (Item #1) or the kW (Item #2) was calculated.

Item #2—Actual Demand

This item (kW) indicates what the customer actual peak demand (966 kW) level was during the billing period (Dec 06 to Jan 03). This figure represents only the customer's peak (kW) not what the utility's peak (kVA) might have been, in this example (1,344 kVA). (See Item #1.)

Item #3—Usage (kWh) Calculations

On this particular billing, usage (kWh) is divided into two different categories (peak and off-peak), and each of these categories has different cost components. (See Item #7.)

Peak periods normally are times when utility usages are relatively high, which for example could be from 8:00 a.m. through 8:00 p.m. on weekdays. If these hours were peak, then all other hours would be off-peak. Every utility that utilizes this type of billing can provide a customer with the specific time periods in each category.

Item #4—Rate

The specific rate schedule under which this rate is calculated is Time-of-Use—Large Power—G-3. While on the surface this may not mean much to a typical customer, it does identify a specific rate section that can be accessed that will fully explain what this rate is and how it is calculated.

The purpose of this example is to show a power factor penalty that is not easily found on the face of the billing, and in such instances, the only help will be in reading the applicable rate schedule. For the most part, complete rate schedules can be accessed on each utility's website. Individual rate schedules generally are not over 3-5 pages in length and are relatively easy to understand.

Item #5—Customer Charge

A rate class is needed to assign this charge ($65.58). Generally this type of charge is what a utility assigns as the cost of meter reading, repair, and other miscellaneous general customer services not itemized on the billing elsewhere.

Item #6—Demand Charge

Here is the calculation (1,209.6 kVA × $8.750 = $10,584.00) for what the customer demand cost for this billing period will be. Notice the (1,209.6 kVA) number shown here is present nowhere else on the billing or is the calculation explained anywhere on the billing.

The only way to determine how the 1,209.6 kVA was calculated would be to have a copy of the applicable rate schedule for Time-of-Use—Large Power—G-3 (Item #3). The portion of this rate schedule, concerning power factor, states as follows:

> **Power Factor Calculation—Power Factor Penalties, if any, shall be calculated as follows:**
>
> - The actual peak kVA established during the billing month shall be multiplied by 90% and this number shall be compared to the actual peak kW demand established during the billing month.
> - The higher of the two numbers (peak – kVA × 90% vs. peak kW) shall be utilized as the billing demand for the billing month.

What this means in plain language is that if 90% of the peak kVA is higher than the actual kW, the 90% kVA figure will be utilized. In this example, the calculation would be as follows:

- tem #1: 1,344 kVA × 90% = 1,209.6 kVA
- Compared to Item #2: 966 kW

In this case, 1,209.6 kVA is larger than 966 kW so the billing demand becomes 1,209.6 kVA. The power factor penalty is the difference between 1,209.6 kVA / 966 kW (1,209.6 – 966 = 243.6 units).

It makes no difference that the units have different nomenclatures (kVA / kW). The difference between the kVA (1,209.6) and the kW (966) becomes the power factor penalty (243.6). Since the billing example does not even address "power factor," where is the penalty calculated and how much is it?

In this example, the power factor penalty is buried in the demand charge (Items #2 and #6). Also, in this example the power factor charge (243.6 units × $8. 750 / unit) is $2,131.50. Where is this figure? It is included in the $10,584.00 total demand charge (Item #6). In other words, if the power factor in this example would have been at least 90%— 966 kW (Item #2) + 90% = 1.073 kVA, there would have been no power factor penalty.

It is important to remember that when power factor efficiency is improved, kVA is reduced and kW does not change. In this example, if the 966 kW would have had a 90% or higher power factor, the peak kVA would have been (966 kW + 90% or higher = 1,073.3 kVA or less). In the case of this example, it would have resulted in a demand charge of (966 kW × $8. 750/unit) = $8,452.50 rather than the $10,584.00 (Item #6) shown in this example.

Item #7—Usage Charge

In this example, usage is separated into two time/charge periods (on-peak/off-peak). These time periods are designated in the appropriate rate schedule both with relation to applicable time periods as well as charges per time period. The rationale for a time-of-use type of rate schedule is to charge for electricity in some relationship to its cost to generate at different times of a 24-hour day, as well as for different days of the week.

As can be seen in this example, the charge differential between on-peak and off-peak is large—$0.03658 kWh/on-peak vs. $0.02447 kWh/off peak. On this type of rate, a customer, to the extent possible, should try to shift as much of their usage (kWh) to the off-peak period as is possible.

Item #8—Fuel Charge

This item compensates the utility for their cost of fuel to run their generation equipment

Item #9—OCA Charge (Oil Cost Adjustment)

This charge is not in all utility billings but it represents the utility's oil cost adjustment (OCA) for fuel for its generation facilities. This charge, if it is present, has to be approved prior to its implementation by the appropriate regulatory agency. This charge is not negotiable.

Item #10—High Voltage Meter Discount/High Voltage Delivery Discount

These items relate to a discount the utility gives if a customer takes service at a *high* voltage level. What the high voltage is can be determined by the applicable rate schedule. Typically, the high voltage level would be any voltage over secondary distribution levels. To qualify for high voltage discounts, the customer would have to have a meter voltage of over, typically, 480 volts. (Ex: 960, 1,200, 2,400, or 4,800 volts, etc.)

To qualify, the customer will have to own/lease the appropriate transformation equipment to change (reduce) the incoming voltage to usable levels (480, 240, 120 volts). In this example, the total discount amounts to $630.99 ($278.11 + $352.88). Whether this is a cost-effective measure to be utilized depends upon what the transformation equipment costs, and whether it is purchased or leased. See Voltage Level section in this publication for more information concerning different voltage levels.

Item #11—Total Current Amount/Total Account Balance

The point that has been illustrated in Figure 4-13 is that power factor charges can exist on a billing and not be specifically listed or itemized.

It is always the customer's responsibility to know what their billing is telling them regardless of the format utilized by the serving utility. Never assume that power factor is not a cost, even if you cannot see any reference to or charge for power factor on the face of the billing.

Always evaluate any charges on a billing in terms of how it could be reduced and/or eliminated. Consider redirecting a billing charge to correcting the reason for the charge. A billing charge almost never accrues any value. Whereas, correcting the reason for the charge can eliminate the charge sometimes in a relatively short period of time.

Definitions

1. **Capacitance**
   The ratio of an impressed charge on a conductor to the corresponding changes in potential. The ratio of the charge of either conductor of a capacitor to the potential difference between the conductors.

2. **Capacitor**
   A device for accumulating and holding a charge of electricity consisting of two equally charged conducting surfaces having opposite signs and separated by a dielectric.

3. **Dielectric**
   A non-conducting substance or insulator. A substance in which an electric field can be maintained with a minimum loss of power.

4. **Induction**
   The process by which a body having electric or magnetic properties produces magnetism, an electric charge, or an electromotive force in one of the currents. Examples of inductive loads: motors, transformers, fluorescent lamp ballasts.

5. **Kilovolt-Ampere (kVA)**
   An electrical unit, equal to 1,000 volt-amperes.

EXAMPLE #1

Power Factor Adjustment. The rate is based upon the operation of the customers equipment at an average monthly power factor of 85%. When the average power factor is above or below 85%, the sum of the demand and energy charges will be multiplied by the following constants:

Average Monthly Power Factor	*Constant*	*Average Monthly Power Factor*	*Constant*
100%	0.9510	70%	1.0835
95%	0.9650	65%	1.1255
90%	0.9810	60%	1.1785
85%	1.0000	55%	1.2455
80%	1.0230	50%	1.3335
75%	1.0500		

EXAMPLE #2

Power Factor Adjustment. When customers power factor is less than 85% lagging, utility may adjust the kW measured to determine the demand by multiplying the measured kW by 85 and dividing by the actual power factor.

EXAMPLE #3

Excess kVA Charge. $0.58 for each kVA of demand in excess of the maximum demand in kW created during the on-peak and off-peak hours. Such kVA shall be determined by dividing the maximum demand by the average power factor for the month.

EXAMPLE #4

Reactive Demand Charge. $0.59 per kVAR over one-third (1/3) of kW demand.

Figure 4-14. Examples of Rate Schedule Provisions Regarding Power Factor Adjustments.

6. **Kilowatt (kW)**
   A unit of power, equal to 1,000 watts.

7. **Kilowatt-hour (kWh)**
   A unit of energy equivalent to the energy transferred or expended in one hour by one kilowatt of power and is approximately 1.34 horsepower-hour.

8. **Reactance**
   The opposition of inductance and capacitance to alternating current, equal to the product of the sine of the angular phase difference between current and voltage and the ratio of the effective voltage to the effective current. A reactive load maintains the magnetic load in motor or transformer ballast.

9. **Resistance**
   A property of a conductor by virtue of which the passage of current is opposed, causing electric energy to be transformed into heat. Examples of resistance loads are electric resistance heat and incandescent lamps.

Figure 4-15 Explanation—(Power Factor)

POWER FACTOR COST:

250	kW Billed
167	kW Actual
83	kW Penalty

- 83 kW × $9.5299 per kW = $790.98 PENALTY
- OR 16% of TOTAL COST LESS TAXES

Power Factor Overview

Items that can cause poor power factor:

1. Oversize/idling motors
2. Lighting ballast transformers
3. Welders
4. Induction heating equipment

IV. VOLTAGE LEVEL

Typical Voltage Levels

- Secondary .... 440 Volts or less
- Primary .... Over 440 Volts
- Savings Potential: ..... 1% to 3%

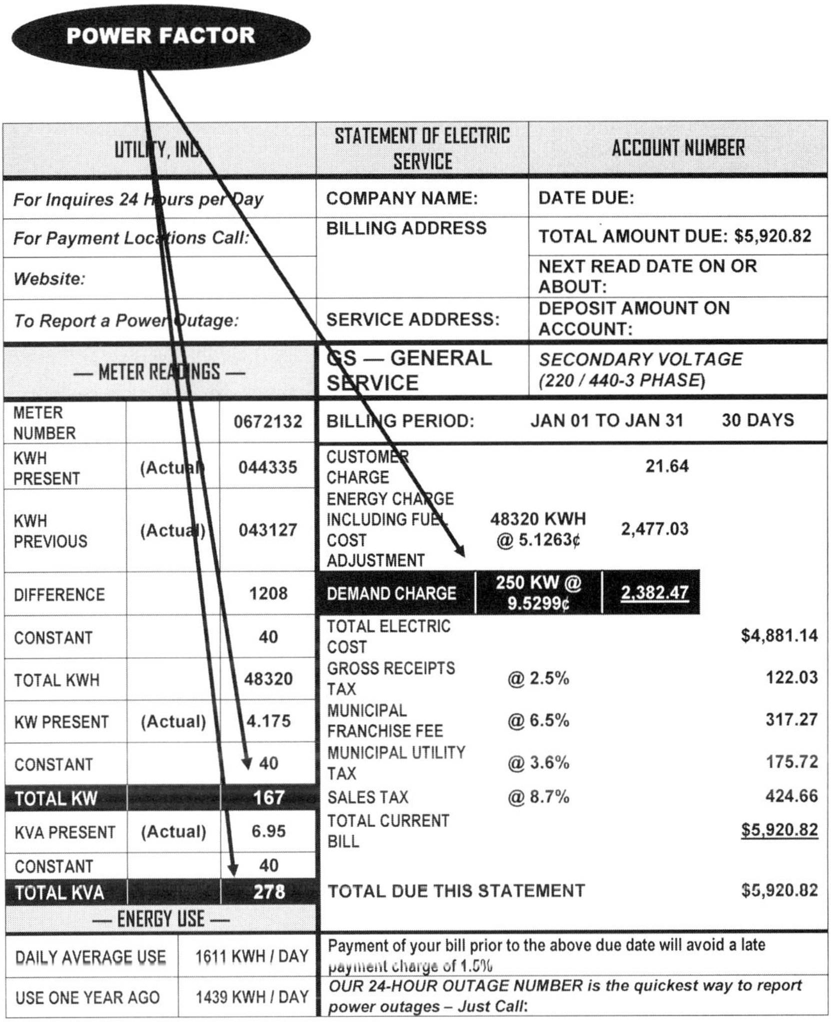

UTILITY, INC.	STATEMENT OF ELECTRIC SERVICE	ACCOUNT NUMBER
For Inquires 24 Hours per Day	COMPANY NAME:	DATE DUE:
For Payment Locations Call:	BILLING ADDRESS	TOTAL AMOUNT DUE: $5,920.82
Website:		NEXT READ DATE ON OR ABOUT:
To Report a Power Outage:	SERVICE ADDRESS:	DEPOSIT AMOUNT ON ACCOUNT:
— METER READINGS —	GS — GENERAL SERVICE	*SECONDARY VOLTAGE (220 / 440-3 PHASE)*

METER NUMBER		0672132
KWH PRESENT	(Actual)	044335
KWH PREVIOUS	(Actual)	043127
DIFFERENCE		1208
CONSTANT		40
TOTAL KWH		48320
KW PRESENT	(Actual)	4.175
CONSTANT		40
TOTAL KW		167
KVA PRESENT	(Actual)	6.95
CONSTANT		40
TOTAL KVA		278

— ENERGY USE —

DAILY AVERAGE USE	1611 KWH / DAY
USE ONE YEAR AGO	1439 KWH / DAY

BILLING PERIOD:	JAN 01 TO JAN 31		30 DAYS
CUSTOMER CHARGE		21.64	
ENERGY CHARGE INCLUDING FUEL COST ADJUSTMENT	48320 KWH @ 5.1263¢	2,477.03	
DEMAND CHARGE	250 KW @ 9.5299¢	2,382.47	
TOTAL ELECTRIC COST			$4,881.14
GROSS RECEIPTS TAX	@ 2.5%		122.03
MUNICIPAL FRANCHISE FEE	@ 6.5%		317.27
MUNICIPAL UTILITY TAX	@ 3.6%		175.72
SALES TAX	@ 8.7%		424.66
TOTAL CURRENT BILL			$5,920.82
TOTAL DUE THIS STATEMENT			$5,920.82

Payment of your bill prior to the above due date will avoid a late payment charge of 1.5%

OUR 24-HOUR OUTAGE NUMBER is the quickest way to report power outages – Just Call:

Figure 4-15. Power Factor—Sample Billing

COST COMPONENTS OF ELECTRICITY—VOLTAGE LEVEL

Many small and medium size industries are taking over a service once provided solely by the electric utility. That service is converting the high voltage power that runs through a utility's transmission lines (128,000; 69,000; 4,100; volts, etc.) to the lower voltage (440, 220, 110 volts) used by most customers.

These industries are doing this by their own purchasing or leasing of transformer installations. When a customer switches to a so-called primary or high voltage rate from a secondary or low voltage rate, generally electric costs are reduced.

Although large industrial users such as steel mills, aluminum processors, large assembly plants, etc. have been purchasing electricity at high voltage rates for many years, smaller industries are now investigating the procedure because of its cost reduction potential.

Also for the most part, utility companies are willing to assist a customer in this effort since it helps them to attract and keep current business. The savings potential is not only for industrial users, but also has as much application for municipalities, universities, shopping centers, hospitals, etc. to name a few.

Can there be significant savings? Typical savings range from 1%-4 % of total electrical costs when comparing primary versus secondary voltage levels of electricity. The cost to convert to primary from secondary service varies greatly but the cost can be minimized when utility-owned transformers are purchased at the depreciated value.

Successful conversions to primary voltage have been possible in instances where the total electric cost was as low as $12,000 per year. This area is one that has great potentiality in electricity cost reduction, plus the process is simple to start.

The utility is your source of information and they probably will be cooperative in helping you to determine your potential for savings. To assist you in a logical step-by-step analysis of the potential for primary voltage in a particular situation, the following guide is given.

If a client is served at secondary voltage level, and if the rate schedule provides for primary voltage discounts, perform the following investigation.

1. Have the utility company calculate the annual savings of switching from secondary to primary voltage levels. (There should be no charge by the utility for doing this analysis.)

2. Have the utility company calculate the depreciated value of the transformation equipment necessary to convert to primary voltage. (Utilities generally depreciate straight-line method in from 30 to 40 years.)

3. Determine the most economical method to convert to primary voltage as follows:
   A. Purchase transformation equipment at depreciated value.
   B. Lease transformation equipment from the utility if applicable.
   C. Lease transformation equipment from a third party if utility lease is not available. A third-party lease is accomplished by purchasing the transformation equipment at its depreciated value from the utility and selling it to the third party that takes title to the equipment. The third party then leases the equipment to the customer just as the utility would do in the utility lease.

	Incremental Steps	*Typical Voltage*
1.	Generation Facility	7,500-13,200 Volts
2.	Voltage Boost Transformation	69,000-750,000 Volts
3.	High Voltage Distribution	69,000-750,000 Volts
4.	Substation Transformation	32,000-69,000 Volts
5.	Primary Transformation	2,400-69,000 Volts
6.	Secondary Transformation	480-110 Volts
7.	Customer Use Point	Meter Point

Figure 4-16. The Incremental Steps in Electricity Distribution for Secondary Voltage Usage Level.

In Figure 4-17, an analysis is shown comparing a customer currently served on a Secondary Voltage Rate schedule with a Primary Voltage Rate schedule. In this analysis *(which is generally provided by the utility* at *no cost),* a potential savings of $194,350 is achieved. The results of this analysis are shown in the figure.

Although the purchase of the transformers involved would seem to be the most cost-beneficial method to utilize (2.57-year payback), the lease method also provides some attractive savings ($110,350/year) as well as other benefits.

If the transformers are purchased, the owner is responsible for all maintenance and has the burden of transformer replacement if required. With the lease option, maintenance and transformer replacements are the responsibility of the lessor, not the lessee.

1.	Potential savings	$194,350
2.	Cost to purchase transformation equipment using current depreciated value of transformers	$500,000
3.	Payback on investment to purchase transformer ($500,000 ÷ $194,350 = 2.57 years)	2.57 years
4.	Cost to lease transformers from a third-party leasing agent yearly, on a 10-year lease, including maintenance and transformer replacement, if required ($7,000/mo; $84,000/yr; $840,000/10 yr)	$84,000
5.	Payback, if transformers leased (yearly) ($194,350 – $84,000 = $110,350)	$110,350

Figure 4-17. Savings Analysis.

Questions to Ask Regarding Secondary and Primary Voltage

1. What is the current voltage level-primary or secondary?
2. Who currently owns the power transformer?
3. If the transformer is utility-owned, what is the depreciated value?
4. Are there any PCB (polychlorinated biphenyl) contamination problems with the transformer that is being considered for lease or purchase?
5. Has an infrared scan been performed on the transformer to locate any hot spots or high temperature differentials that could ultimately result in premature failure?
6. If the transformer were utility-owned, what would the monthly rent or lease cost be?
7. What would a third-party lease cost per month?

Example #1—TRANSFORMER DISCOUNT: $0.055 per month, per kW billing demand to any customer meeting primary service qualifications. The customer must supply their own transformers, substation equipment, etc., and be served at 4,160 volts or higher.

Example #2—DEMAND CHARGE DISCOUNT: If the customer furnishes utility-approved primary voltage transformers, the Demand Charge will be reduced by $0.70 per kW.

Example #3—DISCOUNT: For service at primary voltage, $0.40 per kW demand and $0.07 per kWh.

Example #4—DISCOUNTS: A discount of 6% of the kWh when energy is metered at primary voltage. When customer furnishes and maintains the required substation for service at primary or transmission line voltage, the following discounts will be allowed on each monthly bill:

Monthly Delivery Voltage	Credit per kW of Billing Demand
2,300-12,000	$0.20
13,200 and over	$0.35

Example #5—DEMAND CHARGE:

$13.00 per kW on-peak billing demand, plus.
$4.75 per kW maximum demand, (less than 24 kV) primary service
$3.00 per kW maximum demand, (24 to 41.8 kV) sub-transmission voltage level
$1.90 per kW maximum demand, (120 kV & above) transmission voltage level

Example #6—DEMAND CHARGE:

$6.00 per kW-secondary distribution
$4.10 per kW-primary distribution
$3.00 per kW-transmission line

Example #7—DISCOUNT: A 5% discount of energy charge when the customer owns and maintains, or at the utility's option, leases all transformers and other facilities necessary to take service at the primary or transmission voltage delivered.

Figure 4-18. Seven Examples of Rate Schedule Provisions with Discounts Given for Primary vs. Secondary Voltage Levels.

8. What is the current monthly electrical cost?

9. What would the monthly cost be if voltage were at primary level?

10. What would the payback be?

VOLTAGE LEVEL

UTILITY, INC.	STATEMENT OF ELECTRIC SERVICE	ACCOUNT NUMBER
For Inquires 24 Hours per Day	COMPANY NAME:	DATE DUE:
For Payment Locations Call:	BILLING ADDRESS	TOTAL AMOUNT DUE: $5,920.82
Website:		NEXT READ DATE ON OR ABOUT:
To Report a Power Outage:	SERVICE ADDRESS:	DEPOSIT AMOUNT ON ACCOUNT:
— METER READINGS —	GS — GENERAL SERVICE	*SECONDARY VOLTAGE (220 / 440 – 3 PHASE)*

METER NUMBER		0672132
KWH PRESENT	(Actual)	044335
KWH PREVIOUS	(Actual)	043127
DIFFERENCE		1208
CONSTANT		40
TOTAL KWH		48320
KW PRESENT	(Actual)	4.175
CONSTANT		40
TOTAL KW		167
KVA PRESENT	(Actual)	6.95
CONSTANT		40
TOTAL KVA		278

BILLING PERIOD:	JAN 01 TO JAN 31		30 DAYS
CUSTOMER CHARGE		21.64	
ENERGY CHARGE INCLUDING FUEL COST ADJUSTMENT	48320 KWH @ 5.1263¢	2,477.03	
DEMAND CHARGE	250 KW @ 9.5299¢	2,382.47	
TOTAL ELECTRIC COST			$4,881.14
GROSS RECEIPTS TAX	@ 2.5%		122.03
MUNICIPAL FRANCHISE FEE	@ 6.5%		317.27
MUNICIPAL UTILITY TAX	@ 3.6%		175.72
SALES TAX	@ 8.7%		424.66
TOTAL CURRENT BILL			$5,920.82
TOTAL DUE THIS STATEMENT			$5,920.82

– ENERGY USE–	
DAILY AVERAGE USE	1611 KWH / DAY
USE ONE YEAR AGO	1439 KWH / DAY

Payment of your bill prior to the above due date will avoid a late payment charge of 1.5%

OUR 24-HOUR OUTAGE NUMBER is the quickest way to report power outages – Just Call:

Figure 4-19. Voltage Level—Sample Billing

Figure 4-19 Explanation—(Voltage Level)

For voltage level 220/440 volts—3-phase:

- Is a rate at primary voltage available?
- How much would it save?
- What would be the cost to lease or purchase the required transformers?
- What would payback be?

Voltage Level Overview

1. Determine savings potential
2. Consider transformer lease
3. Consider transformer purchase
4. Calculate savings opportunity

V. USAGE

- Usage is the connected load x hours or usage
  (1 kWh = 1,000 Watts for 1 hour)
- Savings potential: 3% to 10%

COST COMPONENTS OF ELECTRICITY—USAGE

Usage of electricity is measured in kilowatt-hours (kWh) and is *a function of connected electrical load times hours of use.* Usage is not necessarily related to demand, but on a more frequent basis recently, usage costs are being based upon demand levels. Usage costs are directly related to the quantity of equipment operating and the total hours the equipment is used on a monthly basis.

To reduce usage charges, it is necessary to physically reduce usage characteristics. Electricity usage can be reduced by the utilization of energy efficient electricity consuming equipment. Items that can reduce usage charges include the following:

1. Use of solid-state fluorescent lamp ballasts.
2. Replacement of screw-in incandescent bulbs with replacement screw-in fluorescent lamps.
3. Use of high- or low-pressure sodium lamps in areas where color ren-

dition is not critical; e.g., warehouses, parking lots, walkways, etc.

4. Utilization of utility/manufacturer sponsored rebate programs to purchase and install energy efficient electric motors.

EXAMPLE #1

Energy Charge

- $0.04276 per kWh

EXAMPLE #2

Energy Charge

- Ratchet of usage (kWh) to billing demand (kW)
  For first 25 kWh per kW billing demand
  - $0.088125 per kWh first 3,000 kWh
  - $0.043644 per kWh next 87,000 kWh
  - $0.034355 per kWh all additional kWh

EXAMPLE #3

Energy Charge

- Time-of-use/Time-of-day
  —$0.0450 per kWh (off-peak)
  —$0.0668 per kWh (on-peak)

EXAMPLE #4

Energy Charge

- Time-of-use/time-of-day as well as time-of-year

	June-October	November-May
1. On-peak	$0.14605 per kWh	$0.06875 per kWh
2. Intermediate	$0.07632 per kWh	$0.06458 per kWh
3. Off-peak	$0.05947 per kWh	$0.05589 per kWh

Figure 4-20. Examples of Rate Schedule Provisions Concerning Energy Charges (kWh)

Areas to Consider in Reducing Electricity Costs

1. *Negotiation of Equipment Costs*
   - Transformation costs—primary vs. secondary
   - Equipment costs—increase of electrical usage
   - Equipment costs—new construction/new service territory

2. *Municipal and Cooperative Utilities*
   Request pass-through of rates. This rate is offered by utilities that

wholesale to the municipal and cooperative utilities for the purpose of reducing a customer's electricity costs.

3. *Evaluate Demand Effect on the Rate Schedule being Utilized.* Check to see if there is a less expensive rate schedule available if the current demand (kW) is close to a minimum break point. If so, add load to reach break point.

Example:
1) Current Load............... 490 kW
2) Average Cost....................................$.045/kWh
3) Break Point.................. 500 kW
4) Average Cost....................................$.039/kWh

Add load of 10 kW to reach the less expensive rate schedule break point. Also, check to see if utility will allow payment for minimum load (500 kW) without actually reaching (500 kW) peak demand.

4. *Combination vs. Totalization of Demand*
   A. Reduction of total demand (kW). When multiple meters are utilized, demand (kW) is typically totalized (added together). This can result in demand (kW) values being higher than what the utility actually experiences in any single integration period, since separate demand (kW) peaks practically never occur during the same integration period.
   B. Reduction of customer basic charge.

5. *Resetting of Meters*
When a billing (meter) location customer's name changes, the meter should be reset. This is especially true if demand (kW) ratchets are used, since a minimum ratchet demand may be charged whether any usage occurs at all. This is especially important in shopping centers or other places where individual meter point customer names change frequently.

6. *Power Interruptions*
When power interruptions occur due to utility distribution or weather problems that result in electrical outages due to switchgear trip protection devices, do the following things:

1) Record the day the outage occurred
2) Record time-of-day the outage occurred
3) Notify the utility of outage details as soon as possible
4) Analyze the utility bill for the month that electrical outage occurred. If it appears that the electrical outage caused an increase electrical cost, request that the utility adjust the bill and deduct the penalty caused by the electrical outage.

Many times, electric utilities are unaware of all interruptions that occur on their distribution systems. They generally are willing to work with a customer to deduct the cost penalty the interruption caused if they are notified of the interruption promptly. By law, they generally are not required to adjust billings due to these types of interruptions but generally will work with a customer in a situation of this type.

7. *Utilize Onsite Emergency Generation Capacity to Peak Shave Demand (kW/kVA).* Although this process requires parallel wiring of the emergency generation equipment to the customer's general electrical distribution system, in many instances it is a cost-effective measure to consider.

8. *Utilize Onsite Emergency Generation Capacity to Qualify for Utility Demand Reduction Programs.* Some utilities offer programs to customers (with onsite generation capacity) that pay the customers to utilize their generation capacity when asked to by the utility. Also, if a utility offers interruptible electricity rates, the availability of an onsite source of generation can make this type of interruptible rate applicable to a customer at a considerable cost savings.

   The one negative aspect of this process is the fact that the onsite generation source has to be wired in parallel with the onsite electrical distribution system, which can be a costly process.

Explanation on Figure 4-21—Demand Combination Example

Item #1—Rate Classification

This item identifies the customer's applicable rate schedule classification (LCP-1). In this particular instance, the rate classification identifies the type of rate (Large Contract Primary—Category #1). The various kW and kWh charges are determined by the rate schedule provisions.

1	Rate Classification	LCP-1				
	Billing Period	09/01 – 09/30 (30 days)				
	METER READING INFORMATION					
	STATION #1					
2	Meter #	ZX1372				
	Present	42,606	Date	08/31	4	
3	Previous	39,766	Date	09/30		
	Difference	2,840	Meter Constant		300	852,000 kWh
5	Peak Registered Demand	1,563 kW				
6	Totalized Contribution	1,439 kW				
	STATION #2					
2	Meter #	JT17462				
	Present	34,539	Date	08/31	4	
3	Previous	33,577	Date	09/30		
	Difference	962	Meter Constant		750	721,500 kWh
5	Peak Registered Demand	1,590 kW				
6	Totalized Contribution	1,434 kW				
	STATION #3					
2	Meter #	BD92721				
	Present	7,178	Date	08/31	4	
3	Previous	6,999	Date	09/30		
	Difference	962	Meter Constant		600	107,400 kWh
5	Peak Registered Demand	270 kW				
6	Totalized Contribution	179 kW				
	TOTALS					
7	Customer Charge					$100.00
8	kWh Charges	(1,680,900 kWh x $0.02670)				$44,880.03
9	kW Charges	(3,052 kW x $16.00)				$48,832.00
10	Facility Charges					$2,161.02
11	Outdoor Lighting Charges					$166.06
12	**TOTAL CHARGES**					**$93,812.03**

NOTE: This example is from an actual customer utility billing. All of the items represented are as shown on this particular customer's billing detail from their utility.

Figure 4-21. Demand Combination Example

Item #2—Meter Number

The three meters utilized in this billing (ZX1372, JT17462, 8D93721) are identified here.

Item #3—Meter Usage (kWh) Readings

The readings shown as *present* and *previous* in each of the three meter stations represent the actual meter register readings. The present readings indicate the meter register readings on 08/31. The *difference* readings are simply the numeric difference between present and previous meter register readings. Present meter readings become the next billing period's previous reading.

Item #4—Meter Constant/Total kWh

Each individual meter point calculates total usage (kWh) here. The meter reading *difference* number times the individual meter constant equals the total usage (kWh) during the current billing period. Each meter has its own multiplying constant. In this example, Station #1 (300), Station #2 (750), and Station #3 (600).

These numbers are a calculation of the indicated meter reading usage compared to actual usage (kWh) that occurs. Very few meters indicate actual true usage (kWh) directly without the use of a meter-multiplying constant. The reason for this is that it is much more expensive to design, build, and maintain a meter of the size that would be required for a large electricity load on a direct-read basis. As a result, most meters sample on a continuous basis and utilize a correction factor (meter-multiplying constant) to calculate true and actual usage kWh. The actual usage (kWh) by meter station for the billing period is:

Station #1	852,000	kWh
Station #2	721,500	kWh
Station #3	107,400	kWh
Total Usage	**1,680,900**	**kWh**

Item #5—Peak Registered Demand

The peak-registered demand by meter station represents the highest (peak) demand that actually occurred during the current billing period. In this example, the actual peak demands registered by meter stations were:

Station #1	1,563	kW
Station #2	1,590	kW
Station #3	270	kW
Total	3,423	kW

For this total to represent the true total for the billing period, it has to be assumed that each station demand peak would have to occur during the same time interval, something that almost never happens.

Item #6—Totalized (Combined) Contribution

These figures represent the calculated actual totalized (combined) demand for all three-station meter points. This calculation compares all three-station meter point demand levels for the entire billing period and calculates the total highest peak for the billing period irrespective of individual station meter point peaks. In this example, the calculated demand by individual meter point is:

Station #1	1,439	kW
Station #2	1,434	kW
Station #3	179	kW
Total	3,052	kW

The reduction in billing demand can be considerable by calculating the maximum combined station meter point demands without regard to individual station meter point peaks. In this example, the difference is as follows:

Station #	*Peak Registered Demand*	*Totalized Contributions*	*Difference*
#1	1,563 kW	1,439 kW	124 kW
#2	1,590 kW	1,434 kW	156 kW
#3	270 kW	179 kW	91 kW
Totals	3,423 kW	3,052 kW	371 kW

Peak Registered Demand	3,423 kW
Totalized Contributions	– 3,052 kW
Difference	371 kW

Based upon the per kW cost of $16.00 in (Item #9), the cost difference is: 371 kW × $16.00 = $5,936.00

On this sample billing, this $5,936 amounts to over 6% of the total billing amount of $93,712.03: ($5,936 ÷ $93,712.03).

Item #7—Customer Charge

This item represents the minimum charge ($60.00) for which the customer is obligated even if there is no electricity used. Generally, this type of charge includes items like meter maintenance, reading of the meter, processing of the billing, etc. This type of charge is not negotiable and is part of a given rate schedule.

Item #8—kWh Charges

This item shows the total kWh from all three station meter points and calculates the total charge for these kWh. The total kWh charge of $44,880.03 is calculated as follows:

Station Meter Point #1	852,000 kW
Station Meter Point #2	721,500 kW
Station Meter Point #3	107,400 kW
TOTAL	1,680,900 kW

(1,680,900 kWh × $.02670/kWh cost = $44,880.03)

Item #9—kW Charges

This item shows the total combined kW demand charge for the billing period:

Demand Unit Cost	$16.00	
Totatized Contribution Amount	× 3,052	kW
Total Demand Charge	$48,832.00	

Note that even with the advantage of meter combination, the demand charge is greater than the usage charge. (kWh charge $44,880.03 vs. kW charge of $48,832.00). Stated another way, demand charges are over 52% of the total billing charges (48,832.00 ÷ $93,812.03 = 52.1%).

When it is realized that demand provides a customer no energy but simply reserves generation capacity, it becomes apparent how important it is to control demand peaks. (See Demand Section in this manual.)

How much does combination of demand peaks save on this particular billing? If the individual station meter points were totalized, the billing demand would have been:

Station Meter Point #1	1,563	kW
Station Meter Point #2	1,590	kW
Station Meter Point #3	270	kW
Total	3,423	kW

The combined demand was 3,052 kW or a difference of (3,423 kW – 3,052 kW = 371 kW × $16.00/kW = $5,936) or 6.3% savings for this billing period. If a customer has multiple meter points in a common structure or on a contiguous piece of property, and the individual meter points are separately billed, always try to have the utility combine the meter points *on paper* (no actual physical/electrical meter changes).

Will the utility do this? It depends upon the utility and how a particular rate schedule is worded with respect to actual demand and billing demands being calculated.

Can a utility do this? Any utility that utilizes electronic pulse recording metering equipment can provide combination of multiple meter points *on paper* by combining individual meter demand peaks together to determine the single largest demand peak without regard to individual meter demand peaks.

This may sound complicated, but for utilities that have electronic pulse recording metering equipment, the process is rather straightforward. Some utilities have a processing charge for combining meter points *on paper,* but if this charge is present, it is generally reasonable ($10-$100 per billing period, per meter point). Any customer that has individually billed multiple meter points on contiguous property should always try to utilize combined *on paper* demand billing. There is no downside to this strategy from the customer's viewpoint.

Item #10—Facility Charges

These types of charges generally relate to some type of special equipment that the utility is providing that is not included as (standard) in the base rate schedule. This type of charge can include items like transformation equipment, special metering power factor correction capacitors on the customer's side of the meter, etc.

Generally, charges like this represent special and/or different equipment the customer has specifically requested to in some way reduce their ongoing utility charges. A customer mayor may not know what these charges represent, but if there are any questions the customer can always ask their utility representative for a copy of the contract between the parties.

Item #11—Outdoor Lighting Charge

This charge represents some type of special outdoor light that the utility is leasing and/or renting to the customer. Generally, these types of rentals and/or leases include the required equipment maintenance of the equipment and billing period electricity usage charges.

Item #12—Total Charges

This entry totals all charge items for this particular billing period (Items #7, #8, #9, #10 and #11) that totals $93,812.03.

ELECTRICITY ANALYSIS INFORMATION

The following is a list of items that assist in documenting current usage characteristics.

1. *Today's Date*—Identifies the date upon which the electricity work sheet was completed.

2. *Period Being Analyzed*—Identifies the period covered by the billing being analyzed.

3. *Total Cost of Electricity Used*—Identifies Is the total billing amount from the utility including everything on the billing.

4. *Quantity of Electricity Used During This Period*—Identifies Represents the total kWh used (consumed) during the billing period represented by this analysis.

5. *Average Cost of Electricity Per kWh for This Billing Period*—Calculates the total (including all items charged for on the billing) cost per kWh of electricity used/consumed during the billing period.

6. *Current Demand Charge Per kW*—Lists the per unit (kW) cost for electricity demand (capacity reservation charge). Typically, this unit cost will be itemized somewhere on the current billing. If this cost figure is not on the billing, it can be determined from investigation of the applicable utility's rate schedule.

7. *Current Actual Demand (Total kW)*—Is the actual (which can be different from billing demand #8) peak kW demand established by the customer during the billing period. This kW figure should always be present on any billing that has a demand charge component.

8. *Current Billing Demand (Total kW)*—This item may be different from the figure in #7. Billing demand may be calculated from something different than current actual demand. If billing demand is different from the actual demand (#7), make certain that the reason for the difference is investigated.

9. *Current Demand Penalty*—Is used to calculate the cost differential between actual demand cost and billing demand cost. If actual and billing demand quantities are the same, the entry on this line will be "0."

10. *Total Demand Cost*—Is the total (actual/billing) demand cost for the billing period being analyzed.

11. *Current Power Factor Penalty/Cost*—Is the power factor (kW vs. kVA) charge for the billing period. Any number other than "0" in this item indicates a penalty from the utility to the customer for inefficient demand (kW vs. kVA) utilization.

12. *Current Power Factor Percentage*—If power factor (kW vs. kVA) is a factor in the billing calculation, there will be an indication somewhere in the billing format relating to what the actual power factor is. This item, if present, will normally be expressed as some percentage figure less than 100%.

    Example: If a utility had a registered peak kVA of 100 during a billing period and the customer had a registered peak (kW) of 85, the power factor percentage would be (85 kW ÷ 100 kVA = 85%). If a power factor penalty is present on a billing but nothing is indicated in the billing format regarding the percentage or perhaps even the total power factor charges, always consult the applicable utility rate schedule for information concerning power factor, its calculation, charges, etc.

13. *State Tax Charges*—Most, if not all, states assess sales tax on at least some retail customer utility purchases. Some retail customers may not be subject to state sales tax with residential customers being the most commonly exempted class. Many states have specific sales tax exemption processes for certain retail utility users such as manufacturing or industrial customers.

    If state sales tax is included on a utility billing and the customer is in a manufacturer or industrial classification, always check the appropriate state sales tax statutes. These sales tax statutes are generally available over the Internet. If a state has specific sales tax exemption classifications and a customer appears to fit into one of these classifications, this customer should do the following things
    a) Complete the correct state sales tax exemption form
    b) Process the exemption application with the state assigned number
    c) Notify the electric utility of the state sales tax exemption status

14. *Transformation Purchase/Rental/Lease Charge*—In many utilities, there are rate schedule provisions that provide a customer with an option for the purchase, rental, or lease of the transformation equipment utilized for the customer's facility involved. If this option is available, it can reduce the customers overall electricity costs since it may be less costly to purchase, rent, or lease transformation equipment than to pay a more costly incremental kW/kWh rate on a rate schedule. Generally, if a customer purchases, rents, or leases transformation equipment, they will qualify for a less incrementally expensive rate schedule rate classification.

15. *Who Owns Transformation Equipment*—The customer or utility. Depending upon who owns the equipment, the following should be considered:

a) If the customer has purchased, rented, or leased transformation equipment from the utility, it should determine whether the purchase, rent, or lease option is the best that can be negotiated.

b) If the utility owns the transformation equipment, the customer should investigate the applicable rate schedule for purchase, rent, or lease options.

16. *Miscellaneous Charges*—These types of charges win vary by utility, state, county and city, and can include items as follows:
    a) City/county taxes/fees
    b) Fuel cost adjustment charges
    c) Special regulatory agency charges
    d) Local franchise charges
    e) Previous billing amounts due

17. *Does the Operation Utilize Hot Water or Steam in any of the Processes*—There may be an opportunity for the utilization of cogeneration at a facility that utilizes hot water or steam in the processes.

OVERVIEW

Whatever form an analysis of this type ultimately takes, it should be a help, not a hindrance. It should be an aid to more thoroughly understand the utility billing process and the specific items being billed. Properly designed and used, it can be of great value to the overall utility cost reduction process.

CIRCUMSTANCES THAT CAN AFFECT ELECTRICITY COSTS

1. *Mandatory Time-of-Use (TOU)/Time-of-Day (TOD) Rates*—Rates that assign different costs for different times of the day

2. *Longer on Peak Periods Rates*—Rates that assign a higher cost for a longer period of the day

3. *Interruptible Power Usage Rates*—Rates that are less costly if power can be interrupted

4. *Real Time Pricing, No Demand Charges, Frequent Energy Charge Change Rates*—Rates that change energy (kWh) costs as frequently as every minute of every hour of every day
5. *More Demand Ratchets Rates*—Rates that assess demand (kW) charges based upon a past demand level rather than the amount actually used in the current month
6. *Demand/Usage Combination Rates*—Rates that assess usage (kWh) costs based upon the amount of demand (kW) used
7. *Negotiated Rate Schedules*—Special customer-specific rates that are negotiated on a customer-by-customer basis
8. *Combined Metering Rates*—Rates that combine father than totalize demand (kW) usages in multiple meter installations.
9. *Deregulated Rate Structures*—Rates that allow a customer to pick and choose different billing components to arrive at an overall cost
10. *Deregulation of Electricity on a Customer Basis*—The purchasing of electricity from a source other than the local electric utility
11. *Frequent Equipment Credit and Negotiation Considerations*—Utility and third-party negotiation processes where the utility and third party will assist with up-front equipment costs
12. *Opportunity to Institute Shared Savings Projects*—Many providers and installers of electricity usage reduction equipment and processes are willing to assist customers in the initial costs of this equipment and/or processes, which benefits both the provider/installer and the customer.

Lighting Sources In Lumens Per Watt (Efficiency)

	Light Type	Lumens Per Watt
1.	Incandescent	10-25
2.	Tungsten Halogen Incandescent	12-37
3.	Compact Fluorescent	35-70
4.	Mercury Vapor	30-85
5.	Fluorescent	37-85
6.	Metal Halide	80-112
7.	High-pressure Sodium	65-135
8.	Low-pressure Sodium	100-185

Figure 4-22. Various Lighting Sources—Lumens per Watt Comparisons

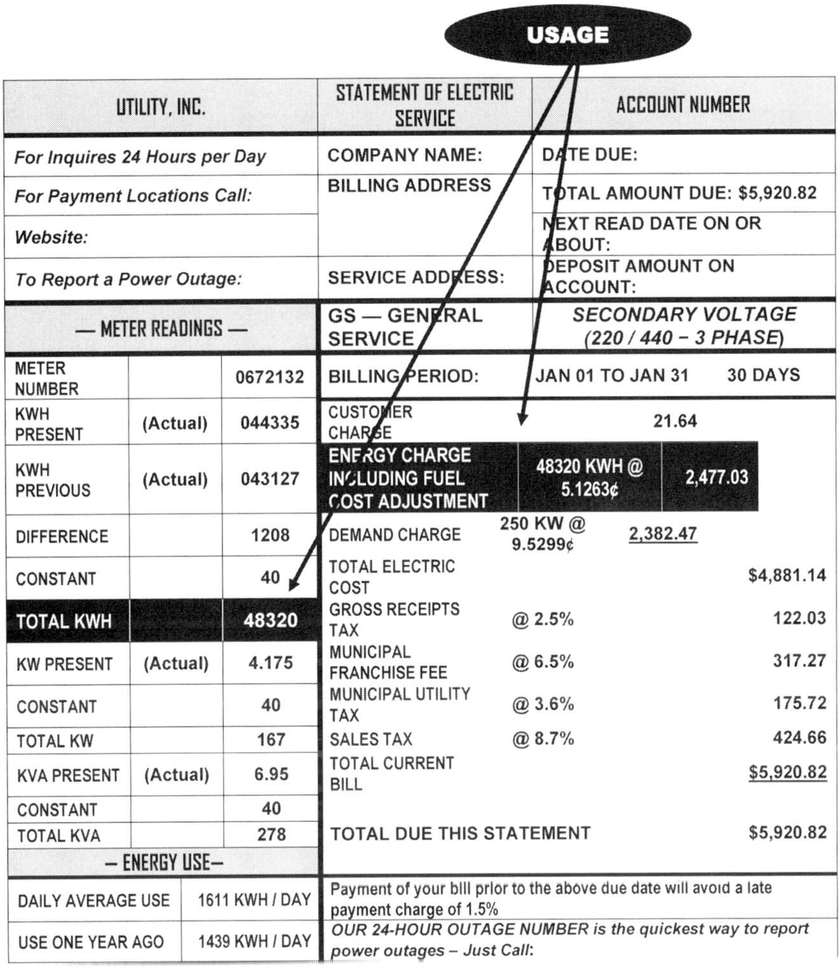

UTILITY, INC.	STATEMENT OF ELECTRIC SERVICE	ACCOUNT NUMBER
For Inquires 24 Hours per Day	COMPANY NAME:	DATE DUE:
For Payment Locations Call:	BILLING ADDRESS	TOTAL AMOUNT DUE: $5,920.82
Website:		NEXT READ DATE ON OR ABOUT:
To Report a Power Outage:	SERVICE ADDRESS:	DEPOSIT AMOUNT ON ACCOUNT:
— METER READINGS —	GS — GENERAL SERVICE	*SECONDARY VOLTAGE (220 / 440 – 3 PHASE)*

METER NUMBER		0672132	BILLING PERIOD:	JAN 01 TO JAN 31	30 DAYS	
KWH PRESENT	(Actual)	044335	CUSTOMER CHARGE		21.64	
KWH PREVIOUS	(Actual)	043127	ENERGY CHARGE INCLUDING FUEL COST ADJUSTMENT	48320 KWH @ 5.1263¢	2,477.03	
DIFFERENCE		1208	DEMAND CHARGE	250 KW @ 9.5299¢	2,382.47	
CONSTANT		40	TOTAL ELECTRIC COST			$4,881.14
TOTAL KWH		48320	GROSS RECEIPTS TAX	@ 2.5%		122.03
KW PRESENT	(Actual)	4.175	MUNICIPAL FRANCHISE FEE	@ 6.5%		317.27
CONSTANT		40	MUNICIPAL UTILITY TAX	@ 3.6%		175.72
TOTAL KW		167	SALES TAX	@ 8.7%		424.66
KVA PRESENT	(Actual)	6.95	TOTAL CURRENT BILL			$5,920.82
CONSTANT		40				
TOTAL KVA		278	TOTAL DUE THIS STATEMENT			$5,920.82
– ENERGY USE–						
DAILY AVERAGE USE	1611 KWH / DAY		Payment of your bill prior to the above due date will avoid a late payment charge of 1.5%			
USE ONE YEAR AGO	1439 KWH / DAY		*OUR 24-HOUR OUTAGE NUMBER is the quickest way to report power outages – Just Call:*			

Figure 4-23. Usage—Sample Billing

Figure 4-23 Explanation—(Usage)

- KWh Charge $2,477.03 (50.7%)
- KW Charge $2,382.47 (48.8%)
- Customer Charge $21.64 (0.5%)
- Total Charge $4,881.14 (100.00%)

- Only 50.7% of the billing total ($4,881.14) was utilized to purchase kWh. Since kWh are the only units that are usable by the customer, 49.3% of the billing total went for other charges. The demand and customer-demand power factor items need to be evaluated to reduce the total billing for this account.

USAGE CONSIDERATIONS

- Analyze facility to determine kWh reduction potential
- Consider energy-efficient equipment
- Evaluate:
  - project design criteria
  - heating system efficiencies
  - motor size and efficiency
  - electronic ballasts
  - occupancy sensors

SYNOPSIS—ELECTRICITY BILLING COMPONENTS

Electricity billing detail is fundamental to the understanding of what the customer is purchasing. If the billing information is evaluated in comparison to the customer's usage characteristics, a cost reduction strategy can be successfully implemented.

Chapter 5

Electricity Deregulation Explained

THE BASICS OF ELECTRICITY DEREGULATION

Deregulation of electricity is simply the purchase of the electricity commodity by retail customers from a source other than their own serving utility. The process is very similar to retail customer purchase of natural gas; however, the effect on the electric utility is very different from that on the natural gas Local Distribution Company (LDC).

In the purchase of electricity from a utility, the commodity (electricity) being purchased is typically considered to be originated or generated by the electric utility. Even though electricity is generally considered to be generated by the utility that sends the monthly bill to the customer, no one really knows what actual electrons flow to the customer's meter point. All electric utilities that generate electricity actually *flow* that electricity into a power pool for ultimate distribution to the customer.

Currently there are many generation areas that cover all of the contiguous United States and parts of Canada. No one can identify or track the actual flow of electrons from the point of generation to the point of use by a customer. Unless there is a specific physical single line that connects electricity generation with the user of that electricity, it is impossible to identify the location where any particular electron was generated, as there are thousands of electricity generation units flowing electrons in the United States.

Electricity flows based upon the path of least resistance, not just where any particular utility wants it to go. There are no good or bad electrons; however, there are various levels of delivery classifications. Whether electricity commodity is purchased from the utility or through the deregulation process, the actual electrons being delivered to the customer will be the same quality and be distributed through the same transmission/distribution wires and meters that are used by the utility that serves the customer.

In the purchase of natural gas, in the majority of instances, the commodity (natural gas) does not originate within the LDC's service territory but is simply purchased by the LDC and resold to the retail customer.

Although the differences between electricity and natural gas may seem to be of little importance to a retail electricity customer, these differences are actually critical to the utilities and their differing attitudes towards the customer's direct purchase of the commodities that they sell.

Since most natural gas LDCs purchase the commodity they sell from someone else, it does not disrupt their operation or profitability to any great extent whether their customers purchase LDC natural gas or arrange for their own natural gas and simply utilize the LDC pipes, meters, and services. If the LDC charges are based upon true cost of service principles, most natural gas LDCs would probably rather retail customers obtain their own natural gas since it would result in less headaches for the LDC.

An electric utility, however, takes a very different view of direct purchase of electricity by its retail customers since generally the serving electric utility, in theory at least, generates a quantity of electricity equal to what it sells to retail customers. When a retail electricity customer purchases the electricity commodity from some source other than the serving utility, the potential lost electricity generation sales for the serving utility result in lost revenue. The lost generation revenue may not be capable of being replaced by the serving utility.

Deregulation of electricity is inherently neither good nor bad, but a case can be made both for and against the process. From a customer's viewpoint, due to the large variation in electric utility rates caused by a lack, in part, of true competition, deregulation of electricity would seem to be a welcomed option.

From an electric utility viewpoint, someone has to pay for its investment in materials and generation capacity; and, if customers can purchase their electricity anywhere, the utility may have no way to recoup its costs.

Deregulation of electricity is not something that is technologically impossible to do since almost all electric utilities currently wholesale electricity between themselves on a daily basis. The real problem electric utilities have with deregulation of electricity is that it requires them to obtain and keep their retail customer base through competition, not through regulatory commission mandated service territory boundaries.

How Deregulation is Evolving

One of the real questions concerning deregulation of electricity is how the electricity gets from the point of generation to the point of use over various transmission lines. The stated problem is one of logistics—how to get the electricity from the point of generation to an individual specific retail customer that is perhaps located in a distribution area different from the one where the electricity is generated.

Questions & Answers about Deregulation

Question: **What about power quality?**

Answer: **Power quality will be what it is.**

The first thing that has to be determined is what is power quality. Power quality means many things to many different people. Power quality can mean that if the power is available it is good quality. To others, it can refer to variation in hertz (cycles), variation in voltage, harmonic levels, and electric line transmitted spikes, etc. Power quality, whatever this means, is consistent both before, as well as after deregulation.

The quality of power is a function of the area the retail customer is located. It must be remembered that generally, retail customers are not directly connected to any single serving utility's generation sources. Retail customers are directly connected to their serving utility's distribution system, but the actual electrons that enter their serving utility's distribution system can be generated anywhere. In fact, the origination point of the electrons may even be from a different area than the one that directly serves the retail customer.

The truth is that the deregulated electricity is the same as what a retail customer is currently receiving from their current electricity supplier. No one can track the actual path electrons take from the point of generation to a given electricity customer unless there is a direct, individual physical connection between the point of generation and the point of use, which is not the case between generators and users today.

Should an individual user require a particular quality level? The best and most effective way of obtaining quality level is to condition the electricity on the user's side of the electricity delivery meter point. Quality of power is not a reason for not using deregulation of electricity.

Question: **How do I know that I will have the electricity I need?**

Answer: **The required electricity will be available if you have opted for firm commodity, transmission, and distribution.**

The real question is are you able to afford the electricity you receive if your marketer defaults (does not deliver) your electricity needs? Generally there is no accurate way to track a specific customer's deregulated electrons from point of generation, transmission, and distribution to the customer's individual meter point on a real-time basis.

The truing up of a specific customer's actual meter point usage on a monthly basis, with what the electricity marketer actually delivered for that specific customer, is accomplished by comparing the serving utility receipt point quantities from the marketer in the customer's behalf for the same period as is the meter point reading period. It is difficult, if not impossible, to compare actual specific customer usages on a real-time basis.

In practice, the total deliveries by the retail customer's electric utility and the usage quantity registered on the retail customer's use point meter are reconciled on a monthly basis. If there are discrepancies, especially deficits, the retail customer has to pay a penalty to the serving utility for system-supply electricity that was required to make up for the electricity marketer's shortfall.

The question is not whether the electricity is available, but rather who pays for any electricity marketer caused problems. In actuality, the retail customer is responsible to the serving utility for any electricity marketer problems, including potential penalties for shortfalls, overages, balancing problems, etc. The way to minimize these potential cost problems is to structure the marketer contract in such a way as to shield or protect the retail customer from the serving utility cost penalties.

Question: **How will I know if I am getting the best deal?**
Answer**:** **It depends upon what you define as the best deal.**

If the best deal is absolutely the least expensive electricity commodity cost available, probably you will never be satisfied that what you are paying is the least expensive that could be had, since the least expensive cost changes every day or even more frequently.

However, if your definition of the best deal is a reasonable cost, with assured delivery from a marketer that can supply all of the assurances that you require, then you can quantify the best deal as what is best for your requirements.

Many times the least expensive initial cost does not translate to the best long-term value. Comfort with the marketer selected, as well as a comprehensive contract that covers all of the variables that can occur, is the only way to assure reasonably cost and satisfactory assurance that the electricity commodity will be available when needed.

Reasonable cost and satisfactory contract terms are a result of knowledge and negotiation with the potential marketer. Luck is not the venue to accomplish either of these goals.

***Question:* Who maintains the customer's meters, other electricity generation, transmission, and distribution infrastructure when the customer utilizes the deregulation process?**
***Answer*: The same entities that currently do.**

All areas of generation, transmission, and distribution of electricity continue to operate with the same federal/state oversight with deregulation as is currently required. Deregulation of electricity only allows a retail electricity customer the option of selecting electricity commodity suppliers, other than their serving utility. Federal/state regulatory oversight of maintenance, quality, and other electricity generation and delivery characteristics remain the same.

What is Deregulation?

- Item involved is the commodity cost only.
- Removal of the commodity cost from the utility billing.
- Requires customer to arrange for commodity purchase.

Commodity Information

- Commodity cost is generally 30-60% of total cost.
- Commodity purchases from marketers do not necessarily reduce costs.

Deregulation Facts

Deregulation does not assure:

- Least expensive energy cost.
- Most efficient energy use.

SYNOPSIS—ELECTRICITY DEREGULATION EXPLAINED

Deregulation is not a complicated process but it differs in each state where it is applicable. Once the process is understood, a customer can utilize its benefits knowing that they are maximizing its savings potential.

Chapter 6

Electricity Deregulation, The Process

CATEGORIES OF DEREGULATED ELECTRICITY

This section addresses the two classes of deregulated electricity that are currently the most frequently utilized. An understanding of these classes of deregulated electricity is necessary to be able to cost effectively purchase electricity from other than the serving utility sources.

Firm Service

Electricity that is purchased under the firm service category is the type that is typically purchased from the serving utility and has the highest priority of delivery. If any electricity is available, it will flow to firm service customers.

Generally, firm service customers do not have backup generation capabilities and as a result, pay the highest rate applicable. Depending upon the serving utility's rate schedules, firm service category customers may- or may not be eligible for other than firm service electricity. If a customer requires an uninterrupted flow of electricity, they will probably purchase firm service distribution capacity from the serving utility; or, they will backup their operations with onsite generation capacity and utilize an interruptible service.

Interruptible Service

This category of service is the type that is the most cost effective to customers. Those customers who can accommodate interruption of electric service on short notice, generally in peak load seasons or situations, will benefit from this class of service.

Interruptible service is less expensive than firm service. A customer who chooses this type of service will probably either have a type of business than can withstand interruption or will have a backup generation source to supplement a disruption of electricity service.

Many times the cost differential between firm and interruptible service is such that a backup generation supply can be obtained with the savings realized. If this can be done, the end result is increased customer flexibility with respect to electricity supply sources.

Interruptible service is becoming more widely available from serving electric utilities even without deregulation in their service territories. Investigate your serving utility's rate schedules to determine the availability and applicability of interruptible electric service to a particular situation.

Which Customers Utilize the Deregulation Process & Are There Other Options Available?

A large number of customers that promote electricity deregulation are large users where electricity costs have considerable impact on their product and processes. Electric utilities are inclined to be responsive to the customers' needs to keep them as satisfied users of the serving utility electricity. These utility efforts are taking many forms to help their customers reduce their electricity cost—e.g. negotiated incremental rates and/or financial assistance to customers.

Not all potential retail customers utilize the deregulation process in their electricity purchase. However, these customers are finding in many cases, that they are able to negotiate less costly electricity rates with their serving utility. As with any situation that involves competition, even those not directly affected in many cases find that they too can benefit from the process.

Figure 6-1 illustrates the various steps that are typically present in the electricity deregulation process. Utilize this data only as a guide to the incremental steps that are present in a deregulated transaction.

Explanation of Figure 6-1, Deregulated Electricity Flow Chart

PRICE POINT #1

1. <u>Electricity Generation Point</u>

This is the generation facility that is supplying the retail customer deregulated electricity. This generation facility has to be physically connected to the electricity transmission system that the customer's serving utility utilizes. The generator may be adjacent to the customer's serving utility or it may be in a different state so long as both parties have access to a common transmission line. The costs accumulated at this point include

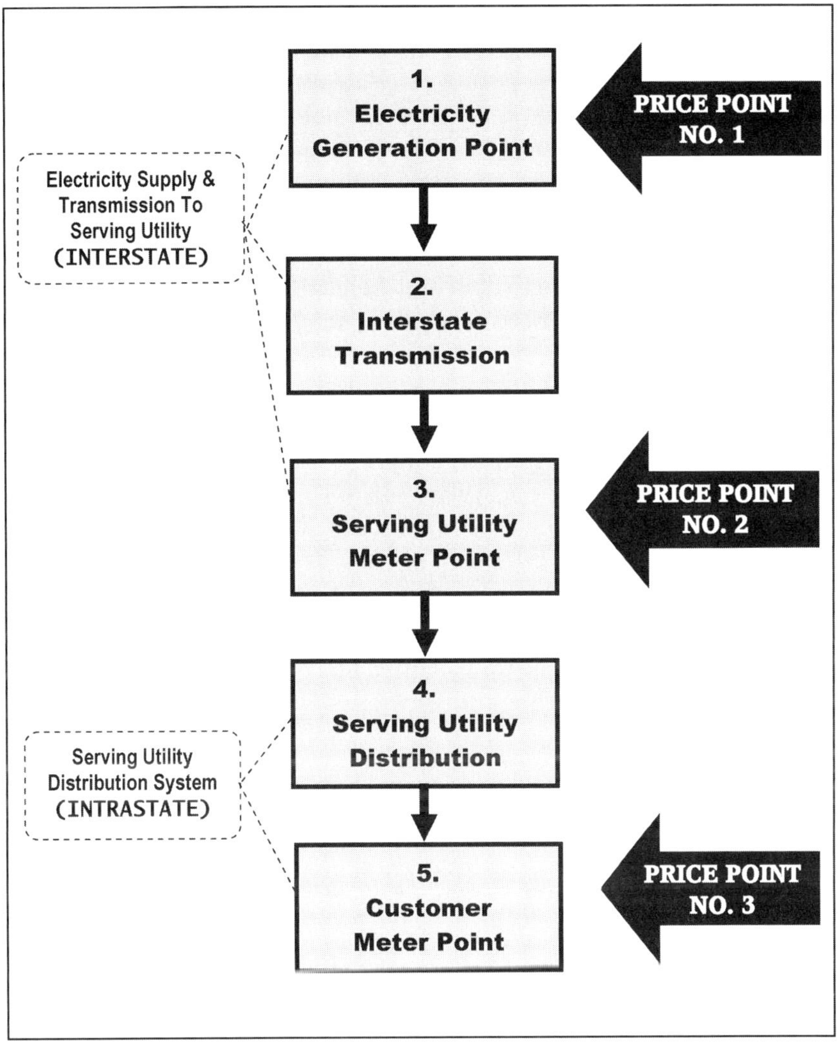

Figure 6-1. Deregulated Electricity Flow Chart

the actual generator charges for the electricity. These costs generally are structured in MWh (1,000,000 watt-hours).

2. Interstate Transmission

This is the transmission line that links the generator to the customer's electric utility. The Federal Energy Regulatory Commission (FERC) governs this process. Generally, the customer has a choice of firm or interruptible transmission service.

PRICE POINT #2

3. <u>Serving Utility Meter Point</u>

This point is where the customer's utility receives, meters, and takes title to the electricity. The costs at this point include the electricity generation costs as accumulated in Price Point #1, the marketer fees, and the interstate transmission costs to deliver the electricity to the customer's utility. Included in the transmission costs are line loss factors, miscellaneous fees, and applicable taxes.

4. <u>Serving Utility Distribution</u>

This portion is the intrastate portion of the system and is part of the customer's serving utility line. It is regulated on an intrastate basis by the appropriate regulatory agencies. In the deregulation of electricity transaction, the intrastate portion of the deregulated transaction is subject to the most regulatory change or deregulation. Line loss factors, miscellaneous fees, and applicable taxes affect this portion of the transaction much the same way as occurred in the transmission line between the electricity generation point and the serving utility meter point.

PRICE POINT #3

5. <u>Customer Meter Point</u>

This is the point at which the electricity passes through the retail customer's onsite electricity meter. The electricity the retail customer actually receives may never include any of the actual electrons that were generated for the retail customer.

The reason for this is that the retail customer's electricity is commingled with all other electricity that is present in the transmission line, as well as between the serving utility meter point and the retail customer's meter point. There is no problem with this since all actual electrons of electricity are the same.

In the deregulated transaction, the actual electrons of electricity that the customer receives are the same as prior to the deregulated arrangement, much like the natural gas transportation process. The deregulated electricity received and metered at the serving utility meter point is recorded and credited to the customer much like a deposit in a bank savings account.

During the billing month, the customer has these deposits available to utilize as determined by the deregulated electricity agreement. The exact electrons deposited in the customer's account may not be the same electrons that are actually utilized. But, so long as there are not more withdrawals than deposits, the overall system will remain in equilibrium.

This may sound complicated, but in fact this is the same thing utilities do daily between themselves on a wholesale basis. It would be very difficult to trace a given electron from generation point to use point, but the system works and remains in balance as long as the same quantity of deposits of electricity are available as there are withdrawals made.

Since electricity cannot be practically stored, this electricity generation and use system must be essentially balanced all of the time. This is no small feat given the complexities of electricity generation and distribution in the United States.

Physically, electricity deregulation works. The problem is in all of the metering and related billing calculations. For example, line losses (under or over), usage of electricity by the customer, and many other cost items that need to be addressed by all of the entities involved. The total deregulated electricity cost to the using customer will be the sum of the costs accumulated in Price Points #1 and #2, and totaled in Price Point #3.

In some scenarios, the customer utilizes interruptible electricity generation and transmission up to the serving utility meter point (Price Point #2) because of the differentials between firm and interruptible electricity costs to these points.

If a customer requires firm or non-interruptible electricity, it may be possible to negotiate a backup arrangement with the serving utility for supplemental electricity in the event of interruption of the customer's electricity. Although deregulation of electricity is nothing like customer transportation of natural gas in technical and operational characteristics, it is similar in the process to the natural gas transportation transaction.

In Figure 6-2 is a typical deregulated cost flow chart that will help to explain the various costs involved in the deregulated electricity process.

Explanation of Figure 6-2, Deregulated Electricity Cost Flow Chart

PRICE POINT #1

1. Electricity Supply

Costs included in this area include the actual commodity (electricity) cost. This cost is calculated in MWh (megawatt-hours—1,000,000 watt-hours). This unit of measure (MWh) is the standard utilized in all power

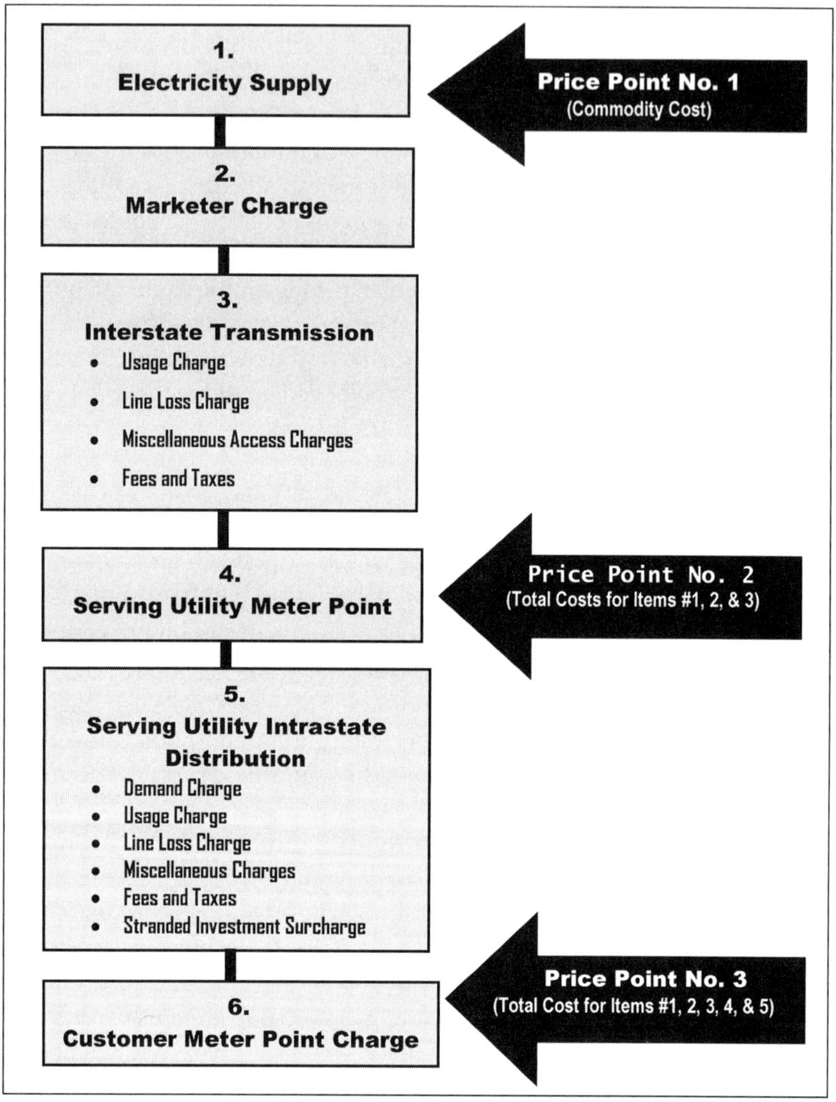

Figure 6-2. Deregulated Electricity Cost Flow Chart

pools for all transactions between utility generators on a wholesale basis.

One of the most readily accessible electricity price indices is the D.J. Electricity Price Index which is found daily in *The Wall Street Journal* in the Money and Investment section. This electricity price data are generally located adjacent to the Futures Price sheets.

The electricity prices in this index represent actual weighted average prices for electricity traded at various delivery points in the power pool regions. Prices are quoted in dollars per megawatt-hour (MWh) and are firm delivery, on-peak and off-peak; and non-firm (interruptible) delivery, on-peak and off-peak.

If any index is utilized, it will only include the commodity (electricity) delivered to a common transmission power pool point. Added to the costs shown in an index will be the transmission costs to the customer's serving utility receipt point, plus the serving utility distribution costs. The various types of electricity that can be purchased at a power pool delivery point are as follows:

- *Firm*
  This class of electricity generation service is expensive, but it is the only type of generation that is available 24 hours a day, 7 days a week with no interruption other than *force majeure* conditions. A customer that cannot be interrupted would opt for this class of service.

- *Interruptible*
  This class of electricity generation service is subject to curtailment based upon the stipulations contained in the contract between the generator and the customer. This class of service is generally less expensive since it can be curtailed, but it can be a very viable option for customers that can interrupt their electricity needs. Generally, interruptions will occur during periods of high electricity usage in the applicable power pool. In most areas of the country, high electricity usage periods occur in the summer months due, at least partially, to air conditioner usage.

- *On-peak Generation Service*
  This class of electricity generation generally can be either firm or interruptible. Generation in this class would occur during the highest usage of generation on the power pool.

  Normally, generation on peak will continue for a period of 12 to 16 hours daily, Monday through Friday. Typically, Saturdays, Sundays, and holidays are considered to be off-peak periods, 24 hours a day.

- *Off-peak Generation Service*
  This class of electricity generation can be either firm or interruptible. Off-peak generation occupies all hours not considered to be on peak. Electricity generation purchased during off peak periods is less expensive than electricity generation purchased on peak. The least expensive electricity generation that can generally be purchased would be off-peak and interruptible.

2. Marketer Charges

This charge represents the customers' cost for the marketer utilized to acquire the electricity generation needed. Most customers utilize the services of a third party (marketer) to initiate and follow-up on the deregulated process much as is generally done in customer transportation of natural gas. These third parties are technically known as agents since they act in the customer's behalf and can legally bind them to a contractual agreement. These entities will perform at least the following functions for the customer:

- Select an appropriate deregulated electricity supply source.
- Select an appropriate interstate transmission line to move the customer's electricity from its origination point to the customer's serving utility meter point.
- Negotiate the least expensive electricity and transmission rates for the customer.
- Assist the customer in the contractual agreements that will be required. There are at least three distinctly different contracts required:
  - Between the electricity supplier and the customer.
  - Between the transmission line utilized, the electricity supply, the serving utility meter point, and the customer.
  - Between the customer's serving electric utility for the transmission and other services that they provide to the customer.

In addition to these (3) contracts, there will be a 4th contract between the customer and the marketer utilized by the customer to facilitate the deregulated process.

Generally, the marketer charge is a percentage of the actual electricity generation cost. Typically, marketer charges are between $0.001 and $1.010 per kWh for electricity delivered to the utility receipt point

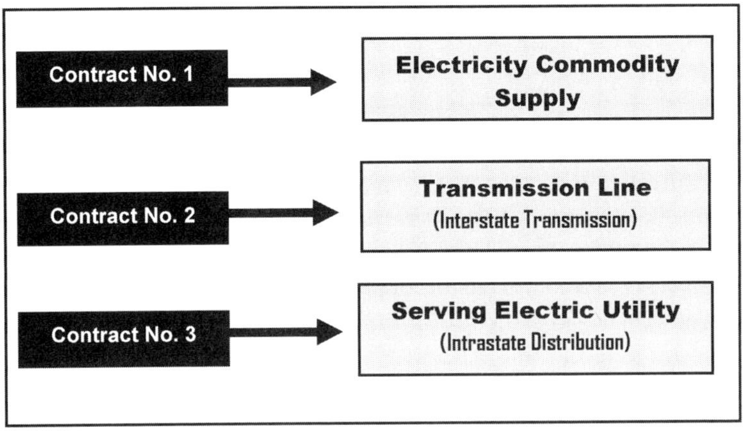

Figure 6-3. Customer Electricity Contracts

3. Interstate Transmission Charges

These changes represent the power pool's transmission cost to move the electricity generated, over the applicable power pool's transmission line, to the customer's serving electric utility metering point. These costs represent rental or fees collected by the various power pool transmission line owners. These fees are subject to FERC approval and generally include line loss fees to account for electricity generation losses due to the resistance of flow of electrons through the transmission system. The loss factors are subject to FERC approval.

To minimize loss, electricity generation voltage levels in this portion of the transaction are generally between +100,000 volts to +385,000 volts. These high voltage levels are generally reduced to more usable levels on the serving utility's distribution system. Also, included in this portion of the transaction would be any miscellaneous

PRICE POINT #2

4. Serving Utility Charges

This is the point at which the marketer exits the deregulated electricity transaction. This is also the point at which the marketer's costs are accumulated and billed to the customer. The billing from the marketer will include the following items:

- Electricity generation cost per MWh (Figure 6-2, Item #1)
- Marketer charge (Figure 6-2, Item #2)

- Interstate transmission (Figure 6-2, Item #3)
- Total cost for Figure 6-2, Items #1, #2, and #3 are generally expressed in dollars per MWh

PRICE POINT #3

5. <u>Serving Utility Intrastate Distribution Charge</u>

This charge represents the serving utility cost to move the electricity received at Price Point #2 (Serving Utility Meter Point) through their distribution system to the customer meter point. This portion of the transaction is under the jurisdiction of the appropriate state utility regulatory authority. Included in this portion of the transaction will be included miscellaneous costs as follows:

- *Distribution line loss fees, if applicable*
  (See Interstate Transmission Charges, Item #3 for an explanation of line loss.)
- *Transformation costs, if applicable*
  These costs are the result of the serving utility's need to reduce the electricity voltage level supplied from the transmission's power pool at serving utility meter point (Figure 6-2, Item #4) These costs may be included in the distribution charge or it may be listed separately.
- *Customer meter point costs, if applicable*
  These costs would compensate the serving utility for the meter, its usage, and maintenance. This cost may be included in with other items or it may be *listed separately.*
- *Stranded investment recovery costs*
  These costs represent the utility's investment in generation infrastructure that may not be utilized by the customer when they purchase electricity through a marketer. These costs are considered to be stranded since the serving utility, in theory at least, may have no other customer to utilize this infrastructure. These costs are approved on a utility-by-utility basis by the appropriate state utility regulatory agency.

  These types of costs will generally be different for each utility and may even be different for different customer classes served by the same utility based upon usage characteristic variations. Also, included in this portion of the transaction would be and state regulatory agency fees and/or any applicable taxes.

  This price point (Figure 6-2) includes all of the serving utility costs as follows:

— Intrastate distribution cost (Item #5)
— Distribution line losses, if applicable (Item #5)
— Transformation costs, if applicable (/tem #5)
— Customer meter costs, if applicable (Item #5)
— Stranded investment recovery (Item #5)
— Any miscellaneous fees and/or costs, if applicable (Item #5)

All of these costs will be billed to the customer by the serving utility. This billing, together with the marketer billing (Figure 6-2, Item #4), will be the total electricity cost to the customer for the entire deregulated transaction (Figure 6-2, Item #6).

6. Total Costs

This total includes all costs from electricity supply point (Item #1), through the serving utility distribution costs (Item #5), and are as follows:

- Electricity supply cost (Item #1)
- Marketer charge (Item #2)
- Interstate transmission costs (Item #3)
- Serving utility distribution costs (Item #5)

The total cost at this point would be compared to the total cost the customer had been paying to the serving utility prior to deregulation.

DEREGULATED ELECTRICITY PROVIDERS

Marketers used for Other than Serving Utility-supplied Electricity

The various marketers used in the deregulated electricity process are detailed in this section. This information provides the background necessary to be able to intelligently determine the type of marketer that should be used to provide the electricity needed and the reliability required. Also the information given herein provides the foundation for doing deregulated transactions.

Deregulation Information

Generally, direct purchase of electricity is provided by at least three different entities, (1) brokers, (2) marketers, and (3) producers. The three entities are explained following:

Brokers

Most deregulated electricity that is available to customers is not obtained from the serving utility. Other parties (marketers) actively market electricity to customers. The main and most important distinction between brokers and producers is the fact that brokers do not take or assume title to the electricity they market.

Brokers act only as third-party facilitators. In fact, they sell for someone else. Brokers perform as marketers, but do not actually take title to the electricity they sell. Either or both the buyer and/or seller pay brokers a fee for their services. This is not to say broker electricity is unreliable or in anyway different from purchasing from a titled source.

Since title does not pass to the brokers, their warranty as to availability is no better than their source provides. In general, the fewer steps required to arrive at the actual electricity generation source, the more reliable the supply. Do not necessarily disregard broker-supplied electricity, but remember brokers are able to provide no better title to the electricity they market than what they have—which is none. If a broker is used, make sure:

- Their source is identified
- The supply is assured for the duration of the contract
- Their source does have title to the electricity that they provide

In general, it is probably better both on long-term cost and reliability, if you contract with either a marketer or an electricity producer.

Marketers

Marketers differ from brokers in that they take title to the electricity they sell the wheeling customer. A marketer takes title to the electricity but does not have or own the generation facilities. Marketers, or marketing affiliates, are also known in the electric industry as "traders."

While all of this may seem confusing, the difference between this category and the broker category is that title to the electricity does pass with the marketer where it does not pass with the broker. The difference between marketers and producers is that producers own electricity generation facilities and marketers do not.

The marketer category is the largest supplier of deregulated electricity. As with any group of individual entities, there are good and bad available. So be certain that any contracts to be negotiated conform in general to the one that is described in this section.

Producers

Producers have title to and own electricity generation facilities. They are the original owners of the electricity and are responsible for its generation and distribution to an interstate transmission line meter point.

Some producers market their own electricity directly to customers. Sometimes several generators join together to form a quasi-cooperative that in turn markets the electricity to customers.

Probably producers as such, are not a dominant force in the electricity market. In general, they sell their electricity production to (1) a transmission line, (2) a serving utility, or (3) a marketer, which in turn supplies the electricity to the customer.

Basic Differences between Brokers, Marketers and Producers

Brokers

Do not take title to or do not own generation facilities

Marketers

Take title to but probably do not own generation facilities

Producers

Have title to and own generation facilities

SYNOPSIS—ELECTRICITY DEREGULATION—THE PROCESS

Generally, there is no reason to limit the choice of the electricity supplier strictly on the basis of the category (Broker, Marketer, or Producer). The criteria for selecting a supplier will be based upon data such as:

- Reliability of supply
- Price
- Transmission distribution routing
- Contract language
- Customer service
- Congeniality between buyer and seller

Making these evaluations is no simple matter. It is possible sometimes to rely upon the broker, marketer, or producer for help. However, the motivation of these three entities might result primarily from a desire to sell their services, not necessarily to satisfy the needs of the customer.

The surest method to follow is to have a set of guidelines like the ones previously described, and evaluate any potential supplier in terms of these guidelines.

Chapter 7

The Electricity Deregulation Contracting Process

THE CONTRACTING PROCESS

The next thing to consider is how to structure a contract with a marketer that is fair for both the marketer as well as the customer. It is important to address the underlying strategies and requirements that will shape the actual contract. Before actually signing any contract, understand what is expected from the arrangement with the marketer.

Following are items that need to be considered prior to the actual signing of any contractual agreement with any marketer. These items are more related to customer characteristics than to electricity characteristics.

For What Length of Time am I Willing to Agree to Partner with a Marketer?

Commodity-only Contracts

These are generally for a time period from 1 year to no more than 5 years, depending upon the customer's tolerance for risk. There is a tradeoff between contract length and cost protection. Generally, the longer the length of the contract, the greater is the commodity cost assuredness.

The downside is that long-term cost assuredness does not necessarily translate into long-term cost minimization. Many things can change in commodity pricing as well as customer requirements and/or usage characteristics. Most customers will tend to negotiate a 1- to 2-year contract when they initially start commodity-only types of arrangements.

Commodity and Engineering Services Contracts

These contracts will require customer commitments for a longer period of time than for commodity-only contracts. Because of engineering services payback periods, contracts for commodity and engineering services will typically be for a time period from 1 to 5 years.

Since commodity and engineering services extend for years, it is important that the customer feels comfortable with the selected marketer. Also, the contract language needs to adequately address any questions that could affect the long-term benefits of the contract.

Determine (within limits) the Cost Reduction that Should Result from the Specific Type of Contract that is Being Utilized

Commodity-only Contracts

Contracts of this type will generally result in cost reductions of between 2% to 4% of the total customer meter point electricity cost, excluding taxes.

Commodity and Engineering Services Contract

Actual cost reductions vary greatly depending upon the energy efficiency of the customer's facilities prior to the inception of the contract. It is not unusual, but not typical, for quantifiable cost reductions in commodity and energy service arrangements to reach over 10% of the total customer meter point cost, excluding taxes.

The Customer has the Ultimate Responsibility for the Value of any Customer/Marketer Contract

This statement may seem unfair, especially since it will probably seem that everyone has more insight and information about the agreement process and content than does the customer. Whether this is true or not, the fact remains that generally the customer will get what they ask or negotiate for. No more and no less!

The Customer Must set the Agenda for Any Potential Marketer Meetings

Potential marketers are always willing to discuss many options and strategies, but the customer will be the primary mover in the process. It will also be the customer's responsibility to document, schedule, and follow-up on the process.

It is best to have a definite method of action planned before ever entering into a dialogue with any potential marketer. Figure 7-1 outlines steps that should be considered by any customer that expects to have a satisfactory relationship with a marketer over time.

The total time that typically elapses between the actual beginning of the process (RFP process to the ending of the procedure) contract/post-

Customer Responsibilities	Potential Marketer Responsibilities
RFP PROCESS	
1. Select list of potential marketers & distribute materials	1. N/A
2. Determine response period with input from potential marketers	2. Work with customer to determine response period
3. Evaluate proposals	3. N/A
4. Contact & discuss specifics with selected marketers	4. Work with customer to arrive at a satisfactory scope of work
5. Select specific marketer	5. Develop scope of work & begin process
Post RFP / Contract Process	
6. Set goals & time parameters	6. Work with customer to develop satisfactory goals & time parameters
7. Negotiation of a contract with the preferred marketer	7. Work with customer to arrive at a mutually agreeable process that can be quantified in a contract
Contract / Post-Contract Process	
8. Evaluate & sign marketer contract	8. Sign contract
9. Begin work as defined in contract	9. Begin work as defined in contract

Figure 7-1. Customer Strategy

contract process can be from 4-12 weeks. While this may seem to be a rather long period of time, there are many items to be determined and many scheduled contacts with the people involved.

Since the time period can be lengthy, it is critical to the success of the process to keep all of the involved parties updated on at least a weekly basis. Again, this responsibility falls to the customer, not the marketer. A time period might be as follows in Figure 7-2. (Some items occur concurrently.)

A successful arrangement between a customer and marketer, whether for an electricity commodity-only contract or for a commodity and engineering services contract is, to a great extent, dependent upon the customer's input to the process.

1.	Develop RFP scope and language	2 Weeks
2.	Select list of potential marketers	1 Week
3.	Send out RFP materials & hold pre-RFP meeting	2 Weeks
4.	RFP response period	2 Weeks
5.	Evaluate RFP responses & select marketer	2 Weeks
6.	Complete contract process	1 Week
7.	Begin work as defined in the contract	2 Weeks

Figure 7-2. Typical Average Time Periods

This list is intended to serve as a general guide to things that any customer should at least consider before doing anything with a potential marketer of commodity and/or services. This list will not answer all questions, but it can raise the right questions that might otherwise have been overlooked.

Questions to Ask Before Starting a Commodity Contract

1. Can you adequately describe what you want in written form?
2. Do you have a project time period defined?
3. Is your supervisor/leader aware of, and in agreement with, the goals and requirements of the project?
4. Is top management of your company in agreement with and supportive of the project—CEO, CFO, Legal, etc.?
5. Do the appropriate support people in your company realize what will be expected of them in this project?
6. When a potential marketer presents a cost reduction strategy, how will you determine if it is realistic?
7. Is cheapest least expensive in the long run? How do you decide this?
8. How long is the proposed potential marketer's contract period? Is the period realistic in terms of cost reductions?
9. When do actual quantifiable savings start? If savings do not start at contract signing, why not? Can the start date be change?
10. Is what the potential marketer receives in the contract realistic in terms of what you receive? Can you measure this? If not, why not?
11. What happens if your utility usage changes—Up? Down? Hours

used? Days Used? Characteristics of use?

12. Are cost reduction guarantees in the contract? If so, can you quantify them? Do your financial people agree with the quantification rationale?
13. Are there early contract opt-outs? If so, are they fair to both parties?
14. What happens if guaranteed savings are not achieved? Who makes up the deficit? Is contract cancelable if guaranteed savings do not materialize?
15. Have your company financial and legal entities been involved in all aspects of the RFP contract process? Do they understand its financial and legal implications?
16. Have all people that will be involved/affected been included during the entire RFP/contract process so they buy into the process?
17. Have you developed an accurate, easily understood tracking process to keep everyone updated on how the cost reductions of the process are progressing?
18. Are you, or is someone in your company, responsible on an ongoing basis for coordinating and substantiating cost reductions as they occur?
19. Is the project coordination and follow-up consuming an inordinate amount of your time? If so, something is wrong with the process.

TYPICAL ELECTRIC CONTRACT PROVISIONS

1 Marketer agreement
2 Contract terms
3 Billing terms
4 Regulatory agency fees
5 Definitions
6 Services provided
7 Type of service
8 Price
9 Billing/Payment
10 Credit check process
11 Late payment penalty
12 Title I Delivery point
13 customer load changes

14 Indemnification
15 Limitation of liability
16 Termination clause
17 Applicable law
18 Parties I Assignment
19 Waivers
20 Notices I Correspondence
21 Confidentiality
22 Signatories

Contract Provisions Explanation

Marketer Agreement
The legal relationship established through this contract.

Contract Terms
Generally, the term is for least a one-year term.

Billing Terms
Marketer billing is generally separate from the serving utility billing.

Regulatory Agency Fees
If fees of this type are assessed, they will be explained here.

Definitions
Any specific terms which need to be explained and understood in the context of the contract.

Services Provided
At this point will be described what the marketer is providing.

Type of Service
There are two types of service available, firm or interruptible.

Price
All details relating to price will be contained in this section.

Billing/Payment
Time the customer has to pay the marketer bill.

Credit Check Process
The marketer's requirement for the customer's credit worthiness will be given at this point.

Late Payment Penalty
If the customer does not pay under the terms described in Item #8, the resulting penalties are described now.

Title/Delivery Point

Point where the marketer relinquishes control of (delivers) the commodity.

Customer Load Changes

During the term of this agreement, if the customer either increases or reduces their commodity requirements beyond some mutually agreed-to point, the procedure to be followed is described in this section.

Indemnification

The procedures each party will take to indemnify and hold harmless the other party in the case of negligence or intentional misconduct by either party.

Limitation of Liability

The limits of liability in the event of either party's nonperformance of any part of this contract.

Termination Clause

When and how termination of this contract occurs.

Applicable Law

The state law that will be utilized if any questions/disputes arise concerning this contract.

Parties/Assignment

The contract requirements of the parties involved should either of them want to assign their interest to a marketer not originally part of this contract.

Waivers

What happens if either party allows a waiver of a default by the other party if this default continues.

Notices/Correspondence

Where, and to whom, are documents from either party to be submitted.

Confidentiality

Disclosure and written consent processes required.

Signatories

The format for both parties' representatives to sign and date the contract.

Questions to Ask Any Potential Supplier

1. Are they registered with the State Regulatory Agency?
2. What is the price per unit? (kWh, Dth, Therm, Mcf, Ccf, etc.)

3. Does the quoted price include all costs?
4. Where is quoted price calculated—utility receipt point, etc.?
5. Is the price fixed or variable?
6. If the price is variable, how often can it change?
7. Are there automatic price changes?
8. Are there any sign up fees?
9. Is there a fee if the contract is terminated early?
10. Are there incentives for signing up?
11. Does the contract expire or rollover (evergreen)?
12. Will there be a separate monthly supplier bill?
13. Will the supplier provide periodic savings data?

Pricing/Contract Considerations

Pricing

1. Fixed price
2. Index price
3. Fixed/index price mix
4. Index/fixed price trigger
5. Discount off utility price
6. Customer specific pricing

Contract

1. Terms
2. Duration
3. Margin/management fee

SYNOPSIS—THE ELECTRICITY DEREGULATION CONTRACTING PROCESS

After reading this chapter, it may seem that the effort required to achieve cost reductions through the deregulation processes is more trouble than it is worth. But as with any new endeavor, initially there is uncertainty.

If the process is learned correctly now, there is a much better chance of being prepared for the future that awaits all electricity users, even residential customers.

Chapter 8

Distributed Generation

HOW DISTRIBUTED GENERATION CAN REDUCE COSTS

This section describes how to save money by using distributed generation. This process is, in broad terms, roughly similar to having an on-site backup energy source. Distributed generation is not a new technology but until recently, it was not a practical option for most customers. The information in this section can be very profitable if properly implemented.

Distributed generation is technically defined as *the sequential production of electrical or mechanical power and useful heat from the* same *primary energy source of fuel.* In general terms, it is the generation of electricity with the resulting heat from the generation process being utilized to make steam or heat water.

The process of distributed generation can turn almost any industrial or commercial concern into its own power generation company. The types of companies that are using distributed generation are very diverse and include:

- Apartments
- Cities (municipals)
- Commercial
- Federal
- Industrial
- Lodging
- Manufacturing
- Paper
- Pharmaceuticals
- Residential
- Restaurants
- Shopping Centers
- State
- Universities

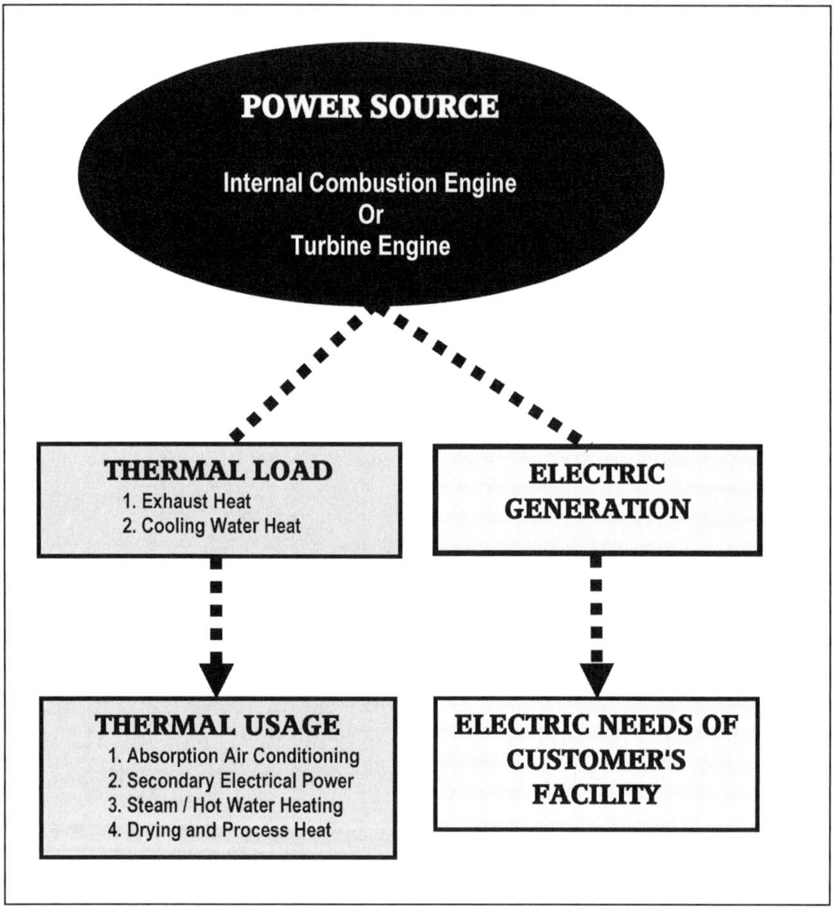

Figure 8-1. Distributed Generation Process

A few years ago, many utilities strongly opposed the customer distributed generation process because they viewed it as a source of competition and a means to erode their monopoly status. Presently utilities now view distributed generation with approval.

This change has been brought about because of the realization on the part of utilities that customer distributed generation can be a way to meet future capacity needs without having to make risky and costly new utility plant investments.

The distributed generation process is an old concept having been made available for many years but only recently been seriously considered as an alternative to electricity rate schedules. This came about because of recent federal energy regulation, primarily the Public Utility Regulatory Policies Act (PURPA). PURPA requires federal rules to stimulate and encourage distributed generation arrangements.

Under PURPA, a distributed generator is allowed to sell the entire output at a price that equals the cost avoided by the utility by not having to generate the purchase electricity. The utility's avoided cost is comprised of two segments, (1) avoided capacity, and (2) avoided energy.

Depending upon the utility, avoided costs can be very low and in the majority of cases will not equal the cost of the distributed generated electricity. In most instances, the use of this process to sell energy to a utility will not be a cost-effective process. The majority of the time, predominate savings in distributed generation will be in the area of use of the process to supply customer electrical power or self-generation. As with any process, proper understanding and implementation are required for a successful outcome to result.

Following are guidelines that will help to determine whether distributed generation or self-generation can be of benefit. Areas that will be explained are:

1. Distributed Generation (self-generation). Is It For You
2. Feasibility Analysis for Distributed Generation (self-generation)
3. Methods of Paying for the Installation
4. Selection of a Distributed Generation (self-generation) Equipment Supplier
5. The Customer/Supplier/Marketer Contract
6. Utility Involvement In Distributed Generation (self-generation)
7. Types of Distributed Generation

If these areas are carefully investigated, then it will be known whether the distributed generation or self-generation process can save money.

DISTRIBUTED GENERATION—IS IT FOR YOU?

Distributed generation is a valuable alterative to the utility electric rate schedule if it is properly used. Generally, to be less costly than utility rates, both components of the process need to be utilized. These components are electricity and heat. The ideal situation for the efficient use of distributed generation is where both electricity and process heat are required.

The process heat requirement generally needs to be in the form of hot water and/or steam. If these requirements are present, then potentially distributed generation can be utilized to reduce electrical costs. If a need for process heat does not exist, it does not necessarily preclude the use of distributed generation but it does raise the cost of the process.

The place to begin in determining whether this process will be of benefit to a given circumstance is to ask the following questions. If the answer is yes to either, then proceed to the next step—the feasibility analysis.

- Are electrical costs more than $.07 per kWh and can process heat be utilized?
- Are electrical costs more than $.09 per kWh but process heat cannot be utilized?

FEASIBILITY ANALYSIS FOR DISTRIBUTED GENERATION

The installation of a distributed generation system begins with the feasibility analysis. Careful consideration at this stage helps to insure the installation of the most cost effective and reliable system applicable. The steps in the feasibility analysis should include the following:

Determination of Heat and Electrical Demand Relationships

This is the most critical step in this process. The areas that are absolutely critical to whether distributed generation will be a viable process are:

- Minimum base daily thermal energy demand
- The coincidence of thermal and electrical energy demand
- The points-of-use and transmission of the distributed generation energy

In this step, determine the heat that can be utilized and it's need in relationship to electrical needs, and the physical relationship of the heat and electrical needs to each other. If this determination is not made accurately, the remainder of the feasibility analysis is of little value. The ability to utilize the thermal off-put from the distributed generation process is the most important and most difficult portion to quantify of the evaluation process.

Determination of Current Electrical Costs vs. Distributed Generation Costs

So that savings information can be developed, the current electrical cost in relationship to the distributed generation costs needs to be quantified in terms of:

- Operation
- Upkeep
- Initial investment

Determination of Energy Cost Savings

Actual energy cost savings are determined by:

- Analyzing cost comparisons relating to distributed generation costs
- The payback from the investment in the generation equipment

This step determines how long it will take a distributed generation system to pay for itself based upon current electrical costs. A typical distributed generation system installed will cost from $800 to $1,800 per kW depending upon the size of the unit and the complexity of the system.

Determination of Equipment Size and Cost

In this step, the initial equipment costs are determined by:

- Thermal demand and the proximity of the locations where energy will be needed
- Proper sizing
- Determination of the number of units required

Coordination of Project with the Local Utility

Coordination of the distributed generation project with the local utility must be executed so that interface problems between the utility and the distributed generation equipment do not occur.

Determination of Physical Equipment Installation Consideration

To be a long-term satisfactory project, the following considerations need to be analyzed for the installation:

- Structural conditions

- Soundproofing needs
- Vibration isolation
- Environmental codes

This step considers the physical aspects of the installation and is one of the most commonly overlooked areas since it is not directly related to distributed generation as such. Any project involving the installation of mechanical equipment needs to be subjected to the same investigation procedures.

When these areas have been analyzed and the findings utilized, it will be evident that the distributed generation system that is specified, based upon the information generated, is the most appropriate and economical one available for a particular circumstance. Do not shortcut the feasibility section of the process since it will pay many dividends in the future satisfaction with the system selected.

Not all distributed generation systems are alike, so carefully analyze how the system will impact operations, capital budget, and the current energy system. Utilize a marketer that is experienced with distributed generation systems, will be able to identify the best alternatives for a facility, and provide confidence about the future operation of the selected system. The marketer utilized also needs to understand:

- System financing alternatives
- Technical specifications of the equipment
- How to integrate a distributed generation installation into a facility

If these points are accomplished, the distributed generation system will be successful.

METHODS OF PAYING FOR THE INSTALLATION

Whether distributed generation is a good investment or not depends upon (1) electrical usage, and (2) current electrical costs.

For example, if the distributed generation process appears to be viable financially with a five-year payback, how is the easiest way to convince a company that the money required should be spent? Obviously, the system can be paid for in several conventional arrangements:

1. Cash

2. Internal company financing
3. Third-party financing
4. Leasing
5. Lease buyback
6. Shared savings

Item #6, *shared savings,* is a rather novel method that is often utilized in the purchase of distributed generation installations. This method is where the supplier of the distributed generation equipment installs and maintains the unit for a specified period of time in return for a percentage of the monthly savings realized by the customer. These savings are determined by the differential between the cost of the utility-supplied electricity and the distributed generation electricity cost.

This method does not require the utilization of up-front capital and provides the customer with ownership of the unit at the expiration of the shared savings period, usually in 5 to 10 years after initial installation. This method of paying for a distributed generation installation is becoming very popular since the customer requires no initial capital investment. Whether or not this method is best depends upon a customer's fiscal policies.

SELECTION OF A DISTRIBUTED GENERATION EQUIPMENT SUPPLIER

As is true of most purchases, the supplier of the service and the equipment is the key to a satisfactory experience with a project of this type. This is especially true in the case of distributed generation since the process and equipment are very faculty-specific in operation and design. When selecting a distributed generation supplier, consider the following things before making a decision:

- Determine how much experience the supplier has in building and installing systems of the type and size needed.

- Obtain a list of current customers, preferably ones that have similar installation, and check out the supplier's performance with these customers.

- Visit an installation similar in design and operation to the one being proposed to be installed so that there will be an understanding of the mechanics of the distributed generation process.

- Based upon information received in Items #1, #2, #3, determine whether the supplier has the experience, abilities, and workmanship the customer requires. If the customer is satisfied with the supplier to this point, proceed to Item #5.

- Determine whether the supplier will perform the feasibility analysis, and if it conforms (in general) to the outline presented in this section. A qualified supplier should be willing to do such a study at no cost.

- Since interface with the local utility is involved, make certain the supplier chosen has experience in the utility's service territory.

- Determine if the contract offered by the supplier conforms, in general, to the one outlined in this section, especially if the project is to be a shared savings arrangement.

If these areas are suitably addressed and answered, then chances of the project being successful are great.

One of the best ways to obtain a list of qualified suppliers is to contact the electric utility company and ask which suppliers have installed facilities on their system of the type being proposed for the application being considered. They can assist in determining specific needs for a given application and generally will provide supplier names with which they have worked in the past in similar installations.

THE CUSTOMER - SUPPLIER CONTRACT

This section will show a typical *Shared Savings Distributed Generation Contract* that is fair to both parties. This outline is to only be used as a guide in any distributed generation installation where a shared savings arrangement is desired. In instances where either the installation is paid for at completion, or typical financing or leasing arrangements are made, normal contracts covering these types of situations will be used.

Figure 8-2.
Distributed Generation Project Development Contract

Distributed Generation
Project Development Contract

Between

____

And

____

Page 1

Figure 8-2 (*Continued*)

INDEX

Page 2

Figure 8-2 (*Continued*)

1.
PARTIES TO THE CONTRACT

THIS CONTRACT, made and entered on this date ____________________
By and Between: ____________________
With Principal Offices at: ____________________
(Hereinafter referred to as SUPPLIER); and,

By: ____________________
With Principal Offices at: ____________________
(Hereinafter referred to as CUSTOMER)

2.
SUPPLIER AND CUSTOMER DECLARATIONS

WITNESS THAT:

WHEREAS, CUSTOMER desires to take advantage of long-term economic savings through the use of distributed generation at its facility; and

WHEREAS, SUPPLIER understands the types and degrees of skills required to develop, own, and operate such systems; and, represents that it has the necessary personnel, experience, competence and legal right to perform such services; and

WHEREAS, CUSTOMER desires to engage SUPPLIER to perform the work and SUPPLIER desires to undertake such performance under the terms, conditions and provisions hereinafter set forth.

NOW, THEREFORE, in consideration of the respective undertakings of the parties, and of the terms and conditions hereinafter set forth, the parties hereto have agreed and do hereby agree as follows:

Page 3

Figure 8-2 (*Continued*)

3.
SUPPLIER RESPONSIBILITIES

SUPPLIER will:

3.1 Feasibility study

Conduct site-specific energy analysis at CUSTOMER'S facility to determine feasibility, annual savings, capital cost, and operating and maintenance costs of an optimal distributed generation system.

*3.2 Savings model

Develop a distributed generation savings model for the specific facility for use in quantifying monthly guaranteed savings achieved. Savings model will be structured to reflect actual minimum guaranteed base electricity and thermal savings.

3.3 Engineering design

Provide detailed engineering system design, equipment installation, electric utility coordination, thermal recovery process, and long-term operation and maintenance of the distributed generation system.

3.4 Communication with CUSTOMER

Communicate regularly with CUSTOMER'S facility personnel, especially during system construction, to plan access to electric and thermal interconnections, and to minimize disruptions to normal plant operations.

*3.5 Correction of malfunctions

Promptly remedy malfunctions upon notification from CUSTOMER.

3.6 Right to discontinue

Retain the right to discontinue project development at any time if, at SUPPLIER'S sole discretion, the project ceases to demonstrate adequate economic value.

(*Mandatory provision in any distributed generation contract.)

Page 4

Figure 8-2 (*Continued*)

3.7 Purchase option

Offer CUSTOMER the option of purchasing the distributed generation system at annual intervals according to a price schedule to be determined before system construction begins.

3.8 Insurance responsibilities

Maintain adequate liability and Workmen's Compensation Insurance coverage while performing work at CUSTOMER'S job site.

3.9 Maintenance responsibilities

Provide required ongoing maintenance for an equipment in the distributed generation system, including replacement of worn parts, change of lubricants and lubricant filters, and periodic adjustments, as required.

4.
CUSTOMER RESPONSIBILITIES

CUSTOMER will:

4.1 Facility accessibility

Allow SUPPLIER and its associates, subcontractors, vendors, consultants, partners, and other interested parties access to CUSTOMER'S facilities for purposes of facility analysis, system construction, operations, and trouble shooting.

4.2 CUSTOMER representative

Appoint a CUSTOMER representative to work with SUPPLIER and its associates to analyze facility energy needs, arrive at the distributed generation savings model, coordinate construction activities and act generally as CUSTOMER'S technical interface throughout the project.

4.3 Access to CUSTOMER data

Provide access to an available previous utility bills (both natural gas and electricity) as well as all information available about plant energy consumption.

Page 5

Figure 8-2 (*Continued*)

4.4Responsibility for utility bills

Continue to be responsible for payment of all utility bills.

*4.5 Shared savings contract

Pay SUPPLIER a monthly fee that wit amount to (__)% of guaranteed real savings attributed to the use of distributed generation in lieu of conventional steam and hot water production and electricity procurement from local utility. The SUPPLIER will pay any impose by the utility in the event the distributed generation unit it is non-operable and the CUSTOMER reverts to utility-supplied electricity. The method of calculating guaranteed real savings will be mutually agreed upon by both parties and will include such items as ongoing utility standby charges, taxes on utility bills, actual electric rates in effect at the time of each calculation, actual natural gas prices in effect, and all other similar items that can be identified as affecting actual savings.

4.6 Natural gas supplies

Use its best efforts to continue to procure natural gas (through marketing arrangements or otherwise) at minimum pricing without risking adequate supply.

4.7 Assurance of operating load

Assure that minimum heat and electric loads remain available at the facility to operate the distributed generation system on a continuous basis, less vacation shutdown periods and maintenance outages. In the event that CUSTOMER cannot provide the energy loads for any reason except an act of God, strike, lockout, war, civil disturbance, explosion, breakage, accident to machinery, failure of distributed generation fuel supply, prevention by federal, state, or local law, binding order of a Court or governmental agency during the term of this Contract, CUSTOMER will purchase system from SUPPLIER according to the price schedule detailed in Section 3.7, or will agree to pay a monthly fee to SUPPLIER for the duration of the load outage equal to the average of the previous twelve (12) month's fee as calculated in accordance with Section 4.5.

4.8 Confidentiality

Maintain confidentiality with regard to this project, and agree not to conduct negotiations with other distributed generation project developers, equipment vendors, energy service companies or other entities until notified in writing by SUPPLIER that SUPPLIER is discontinuing project development due to causes enumerated in Section 3.6.

(*Mandatory provision in any distributed generation contract.)

Page 6

Figure 8-2 (*Continued*)

4.9 Payment of SUPPLIER expenses

Agree to pay SUPPLIER for all out-of-pocket development expenses (including salaries, travel expenses, fringe benefits, etc.) plus a mutually agreed-to cancellation fee if CUSTOMER decides to terminate project development work prior to equipment procurement and installation. If CUSTOMER terminates project development after equipment procurement and/or installation, all associated costs for these functions will be borne by CUSTOMER also (in addition to project development expenses noted above). In lieu of the cancellation fee, if CUSTOMER decides to terminate project due to an electric rate reduction from its electric utility, CUSTOMER will pay SUPPLIER an additional fee equal to 25% of the savings realized by this rate reduction for one full year after the reduced rate goes into effect.

4.10 Water supply and treatment

Provide adequate water supply and water treatment for steam generator makeup.

5.
SUPPLIER INDEMNIFICATION

SUPPLIER shall indemnify, protect and hold CUSTOMER, its directors, officers and employees harmless from and against an loss, costs, damage, injury or expense (including court costs and reasonable attorneys' fees) by reason of any accident, personal injuries, deaths, or damage to property of whatever kind or nature brought by any person, association, or corporation, which loss, damage, or expense is caused by the negligence of SUPPLIER in performing work pursuant to the Contract.

6.
CUSTOMER INDEMNIFICATION RESPONSIBILITY

CUSTOMER shall indemnify, protect and hold SUPPLIER, its directors, officers and employees harmless from and against all loss, cost, damage, injury or expense (including court costs and reasonable attorneys' fees) by reason of any accident, personal injuries, deaths, or damage to property of whatever kind of nature brought by any person, association, or corporation, which loss, damage, or expense is caused by the negligence of CUSTOMER.

Page 7

Figure 8-2 (*Continued*)

7.
INDIRECT OR CONSEQUENTIAL DAMAGES

CUSTOMER shall not be liable to SUPPLIER, and SUPPLIER shall not be liable to CUSTOMER for any special, indirect or consequential damages, including, without limitation, loss of profit, toss of product, and loss of use, arising out of the performance of this Agreement irrespective of either party's fault or negligence.

8.
DELAYS

Any delays in or failure of performance by either party hereto of its duties hereunder (other than the payment of money), shall not constitute default or give rise to any claims for damages if and to the extent such delays or failure of performance are caused by occurrences beyond the control of the party involved. Included are acts of God or public enemy, expropriation of facilities, compliance with any law, or proclamation, regulation, ordinance or instruction of any government or unit thereof, having or asserting jurisdiction, acts of war, rebellion or sabotage or damage resulting there from fires, floods, explosions, accidents, riots or strikes, delay by vendors and delivery of materials and equipment, delay of construction contractors in performing construction work.

9.
NOTICE METHOD

All notices pertaining to this Contract shall be in writing, and if to CUSTOMER, shall be sufficient if sent via first class mail to CUSTOMER at the following address:

Name: ______________________________
Title: ______________________________
Address: ______________________________

And if to SUPPLIER shall be sufficient if sent first class mail to SUPPLIER at the following address:
Name: ______________________________
Title: ______________________________
Address: ______________________________

Page 8

Figure 8-2 (*Continued*)

10.
BINDING OF PARTIES TO CONTRACT

This contract shall be binding upon and inure to the benefit of the successors and assigns of each of the parties hereto.

11.
CONFIDENTIALITY

Any drawings, documentation, specifications, prints, designs, ideas or other information provided by SUPPLIER to CUSTOMER pertaining to the work to be performed hereunder are strictly confidential and proprietary to SUPPLIER. CUSTOMER shall not, without the prior written consent of SUPPLIER disclose any such information to a third party or use any such information for its own benefit other than in connection with the operation of this distributed generation facility.

12.
UNENFORCEABILITY

SUPPLIER and CUSTOMER agree that if any term or provision of this Contract is held by any court to be illegal or unenforceable, the remaining terms, provisions, rights and obligations shall not be affected and shall remain in full force and effect.

13.
ENTIRETY OF CONTRACT

This Contract sets forth the entire understanding of the parties and supersedes all prior contracts, communications, representations or warranties, whether oral or written, by any officer, employee or representative of either party. Any change in the terms and conditions of this Contract must be in writing and signed by both parties.

Page 9

Figure 8-2 (*Continued*)

14.
EXECUTION OF CONTRACT

IN WITNESS WHEREOF, the parties have caused this Contract to be executed by their duly authorized officers as of the day and year first above written.

CUSTOMER	**SUPPLIER**
By: ____________________	By: ____________________
Title: __________________	Title: __________________
WITNESS	**WITNESS**
By: ____________________	By: ____________________
Title: __________________	Title: __________________

Page 10

EXPLANATION OF FIGURE 8-2, DISTRIBUTED GENERATION PROJECT DEVELOPMENT CONTRACT TERMS

1. **Parties to the Contract**
   This section details the parties to the contract, the effective date and the address of both the supplier and customer.

2. **Supplier's and Customer's Declarations**
   This section outlines the reasons for the contract from the prospective of the supplier and customer.

3. **Supplier Responsibilities**
   This section details the responsibilities and rights of the supplier.

   3.1 *Feasibility study*
   This paragraph details the responsibility for and specifics of the feasibility study. This should always be a part of the supplier's responsibility.

   3.2 *Savings model*
   This paragraph outlines the model that will be developed to allow the customer to justify the installation based upon measurable data. It is important that the savings model is based upon actual data, not upon average operational conditions.

   3.3 *Engineering design*
   Here is where the supplier details the system design installation and maintenance requirements of the system.

   3.4 *Communication with the customer*
   This paragraph details the supplier's responsibility to keep the customer informed of the project status. It also notes that the supplier is obligated to minimize customer plant disruptions. This paragraph is general in nature, but it should be in any contract to at least recognize the need for communication and minimum disruptions.

   3.5 *Correction of malfunctions*
   Here is detailed the responsibilities of both the supplier and the

customer in the event of problems. This section must be specific and list all malfunction criteria and corrective actions.

3.6 *Right to discontinue*
This section outlines the supplier's right to stop the project if it ceases to be cost effective. Since this is a shared savings contract, the supplier has to determine if there is adequate potential for profit based upon the shared savings process.

3.7 *Purchase option*
This section gives the customer the option, at predetermined intervals, to purchase the system outright. Some customer may wish to start the project as a shared savings contract because of their lack of knowledge concerning a system such as this. However, once the system is in operation and performing as intended, the customer may want to change to an outright purchase prior to the expiration of the shared savings period. This section provides for this option.

3.8 *Insurance responsibility*
This paragraph assures that adequate insurance is provided by the supplier. If specific insurance coverage is required by the customer's company, the details concerning that insurance would be inserted in this section.

3.9 *Maintenance responsibility*
This paragraph is important especially since this is a shared cost contract. It is import that all aspects of the installation are covered in this section.

4. **Customer Responsibilities**
This section details the responsibilities and rights of the customer.

4.1 *Facility accessibility*
Outlined here is the customer's contract to allow the supplier and various other interested parties to have access to the premises during the installation process. Although this may seem to not be required in detailed form, since it is obvious that the sup-

plier would have to enter the customer's facilities to install the unit, it is well to detail who will be involved in the installation. Sometimes limitations are imposed by the customer because of union contract stipulations or company policy.

4.2 *Customer representative*

This paragraph is important since it is essential to the timely completion of the project to have a customer representative available to coordinate and provide assistance in various portions of the project. This provision can be one of the most important in terms of the customer obtaining a satisfactory installation. Do not skimp in this area. Provide a qualified representative and allocate enough of this time to the project to provide the interface necessary.

4.3 *Access to customer data*

This provision is necessary to allow the supplier to calculate the financial and usage data necessary in determining the viability of the project. Also, these data are required to do the feasibility study.

4.4. *Responsibility for utility bills*

This provision details the customer's continued responsibility for payment of utility bills.

4.5 *Shared savings contract*

This portion of the contract is the heart of the shared savings arrangement. Contained in it is the monthly fee or supplier's portion of the shared savings. The percentage figure will vary depending upon the details surrounding the installation. The percentage share of each party will be predicated primarily upon data obtained during the feasibility phase of the project analysis.

Also, outlined in the *Shared Savings Contract* are the various components that will be considered when making the calculation. The percentage share ratio between supplier and customer will include consideration of all items detailed in this portion of the contract, together with supplier internal calculations including—the supplier's cost of the unit, return on investment requirements, and other cost-return data. It is important that provisions covering non-operable distributed generation conditions do not create penalties that offset savings.

4.6 *Natural gas supplies*
Outlined here is the requirement to utilize the least costly source of energy possible without undue risk to supply considerations. Again, it is important to the shared savings arrangement that distributed generation costs be minimized insofar as is possible so that a payback criterion remains valid and the project viable. If this were a typical purchase or lease contract, then this section would probably not be included since it would affect only the customer if a less cost-effective fuel were used.

4.7 *Assurance of operational load*
This provision assures the supplier that the customer will not cease or reduce usage levels to a point that the distributed generation equipment cannot be utilized. If the customer were allowed to not utilize the equipment, then the supplier's share of the savings would be diminished. If this were not a shared savings contact, this provision would not be included.

4.8 *Confidentiality*
Most suppliers want this type of clause in the contract to assure them that they are not being used simply as a comparison to another supplier. The customer always has the right to cancel the contract subject to the provisions in 4.9.

4.9 *Payment of supplier expenses*
This provision outlines three different conditions of cancellation by the customer:

1. Cancellation prior to equipment purchase and installation
2. Cancellation after equipment purchase and/or installation
3. Cancellation because of utility rate reduction

These provisions protect the supplier if the customer abandons the project through no fault to the supplier. Some of the provisions that are listed under the customer's responsibilities may seem overly protective of the supplier, but remember the supplier is taking all of the up-front risks with equipment and installation costs, which can run into millions of dollars. Therefore, there needs to be some assurance that a capricious act of the customer does not cause the supplier financial harm.

4.10 *Water supply and treatment*
This provision simply states that the water quantity and quality necessary heat load utilization will be available.

5. **Supplier Indemnification**
This section details the supplier's responsibility to protect the customer from any loss, damages, etc. caused by the negligence of the supplier while performing work pursuant to the contract.

6. **Customer Indemnification Responsibility**
This section details the customer's responsibility to protect the supplier from any loss, damages, etc. caused by negligence of the customer while performing work pursuant to the contract.

7. **Indirect or Consequential Damages**
This section details the limits of liability of the supplier and the customer with respect to indirect or consequential damages which might arise as a result of the performance of this contract. The damages included in this section relate only to monetary considerations.

8. **Delays**
This section details those delays that, should they occur, will not constitute default irrespective of the party involved. This includes those items over which neither the supplier or the customer can be reasonably expected to be able to control or foresee so long as due care is exercised.

9. **Notice Method**
Outlined here is the method to be used by either party when providing notice to the other party.

10. **Binding of Parties to Contract**
This section protects both supplier and customer in the event that

either company is sold or absorbed by another company. If this happens, then the successor company is obligated by the terms of the contract to the same extent as was the original party.

11. **Confidentiality**
    This section protects any proprietary data provided by the supplier to the customer in connection with this contract.

12. **Unenforceability**
    If any section of the contract is or becomes legally invalid or unenforceable, it does not negate the remainder of the contract. This clause protects both parties against unenforceable stipulations.

13. **Entirety of Contract**
    This section states that the contract as signed by the supplier and customer is complete as written with no unattached side clause or conditions.

14. **Execution of Contract**
    This section identifies both parties by company and name. The signatures of the responsible parties are notarized in the individual locations of the supplier and customer. When this page is properly signed and notarized by both parties, the contract becomes binding.

UTILITY INVOLVEMENT IN DISTRIBUTED GENERATION

Utility involvement in a distributed generation project is mandatory and close coordination with the local electric utility is required. When the point is reached where a distributed generation contract with the supplier is pending, it is time to contact the utility and coordinate the various interface connections necessary between your equipment and that of the utility.

Although some utilities encourage distributed generation, many will try to dissuade you from installing the system. The reasons given for not doing the project will vary with the utility but in general will revolve around the potential for interface problems between each set of equipment. It is true that an improperly installed system could potentially cause problems; but if the supplier selected is competent, this will not be a cause for concern.

The basic reason for a utility's reluctance to have its customers utilize distributed generation is simple—lost revenue. If the utility cannot convince a customer to cancel a project, they may offer that customer a *special* electric rate to not develop a distributed generation process.

Depending upon the utility and the customer's negotiation skills, this *special* rate can match the distributed generation cost for electricity thermal load, or it can be somewhere between the current utility rate and the distributed generation cost.

Always be aware of stipulations that might be connected with the special rates suggested by the utility. These special rate stipulations may involve long-term (5-15 year) commitments to remain a utility customer.

Some utilities may require that a sum of money equal to the cost of the distributed generation system be deposited with them, at interest, for a specified number of years. This procedure assures them that the special rate the customer received, required customer financial investment equal to the extent of the distributed generation system.

If the method of payment for the system were the shared savings arrangement, then obviously the depositing of funds equal to the system cost is not going to appeal to a customer. A disincentive to a customer is the fact that the rate reduction that the utility will offer will probably have a time limit attached to it so that at some point in the future it will revert to the then current rate structure.

In general, unless a rate reduction can be obtained without unduly restricting the customer's options, it is probably best to proceed with the distributed generation project. Also, because of the engineering design time that will be required before contact with the utility occurs, the supplier will have to be paid cancellation charges if the project is terminated as outlined in the contract between the parties.

TYPES OF DISTRIBUTED GENERATION

There are many types of distributed generation processes insofar as the fuel source and method of generating the electrical power is concerned. Several types of processes are described following:

Internal Combustion Engine-driven Generator

This type of unit utilizes an internal combustion engine that can be fueled by natural gas, propane, fuel oil or gasoline.

Turbine Engine-drive Generator

This process utilizes a turbine engine similar to those used on aircraft to power the generator. It can be fueled by natural gas, propane, or fuel oil.

Steam Boiler Turbine Powered Generator

This process utilizes a fuel-fired boiler to generate steam to power a turbine to generate electricity. The boiler fuel can be natural gas, propane, fuel oil, or coal.

BASIC DATA REQUIRED FOR DISTRIBUTED GENERATION TO WORK

Following is a listing of the basic data that initially needs to be established by the parties to develop a true cost and payback relationship for distributed generation (self-generation).

Facility Data

- Total kWh used per year
- Average demand in kW per hour
- Btu per hour required
- Therms per year required

Distributed Generation Data

- Operating hours
- Total kWh generated per year
- Average load in kW
- Fuel cost:
  - —Diesel fuel
  - —Natural gas
  - —Other fuel
- Heat load available in Btu per hour

Distributed Generation Unit Data

- Unit capacity in kW
- Engine heat output in Btu per hour
- Cost lease amount
- Interest rate
- Term (financing period in years)

Operating Expense per Year
- Debt service cost
- Miscellaneous utility interconnect cost (if applicable)
- Insurance
- Fuel cost
- Operation and maintenance expenses

Plant Revenues
- Cost of electricity
- Value of thermal energy (the actual current cost)

Gross Operating Profit
- Customer's minimum guaranteed savings level
- Installer's percentage share of savings above customer guaranteed level

Calculated Data
- Return on investment
- Simple payback
- Cost per kWh

SYNOPSIS—DISTRIBUTED GENERATION

By the time that you have read over all of the material relating to distributed generation, you may feel that the process is far too complex to even consider, let alone actually installing it at a particular facility. While it is true that the distributed generation process is complex and can be subject to many problems, it is also true that a distributed generation system can be a very valuable energy cost-reduction tool.

To not even consider distributed generation, especially if there is a need for process steam or hot water, is to perhaps overlook one of the most effective energy cost-reduction strategies that may be available to a company.

Distributed generation is a tool that when used wisely can result in very large energy cost reductions. Study the information in this publication incrementally and find that the distributed generation process will become much more understandable.

Chapter 9

Green Power

WHAT IS GREEN POWER?

Green power refers to renewable energy sources that represent those renewable energy resources and technologies that provide the highest environmental benefit. EPA defines green power as *electricity produced from solar, wind, geothermal, biogas, biomass, and low-impact small hydroelectric sources.*

Green power sources produce electricity with an environmental profile superior to conventional power technologies and produce no *anthropogenic* (human caused) greenhouse gas emissions. EPA requires that green power sources must also have been built since the beginning of the voluntary market (1/1/1997) in order to support new renewable energy development.

RENEWABLE ENERGY

Renewable energy includes resources that rely on fuel sources that restore themselves over short periods of time, and do not diminish. Such fuel sources include the sun, wind, moving water, organic plant and waste material (biomass), and the earth's heat (geothermal). Although the impacts are small, some renewable energy technologies have an impact on the environment.

For example, large hydroelectric resources can have environmental trade-offs associated with issues such as fisheries and land use.

CONVENTIONAL POWER

Conventional power includes the combustion of fossil fuels (coal, natural gas, and oil) and the nuclear fission of uranium. Fossil fuels have environmental costs from mining, drilling, or extraction, and emit greenhouse gasses and air pollution during combustion. Although nuclear power generation emits no greenhouse gases during power generation, it does require mining, extraction, and long-term, radioactive waste storage.

TYPE OF EMISSIONS

Human Activity (Anthropogenic) Emissions

Anthropogenic emissions are produced as a result of human activity that unnaturally releases CO_2 emissions into the atmosphere. One of the largest sources of anthropogenic CO_2 emissions is the combustion of fossil fuels or fossil fuel-based products to produce electricity.

Biogenic Emissions

Biogenic emissions, in contrast, result from natural biological processes, such as the decomposition of combustion of vegetative matter. Biogenic emissions are part of a closed carbon loop. Biogenic CO_2 emissions are balanced by the natural uptake of CO_2 by growing vegetation, resulting in a net zero contribution of CO_2 emissions to the atmosphere.

Examples of biogenic emission sources include burning vegetation (biomass) to produce electricity or using plant-based biofuels for transport.

GREEN POWER MARKETS

Mandatory Markets

Mandatory markets exist because of policy decisions, such as state renewable portfolio standards (RPS). Such standards require electric service providers to have a minimum amount of renewable energy in their electricity supply. Often, these policy decisions specify eligible energy resources or technologies and describe how electricity service providers must comply.

Voluntary Markets

Voluntary markets, also referred to as green power markets, are driven by consumer preference. Voluntary markets allow a consumer to choose to do more than policy decisions require and reduce the environmental impact of their electricity use.

Voluntary green power products must offer a significant benefit and value to buyers to be successful. Benefits can include zero anthropogenic greenhouse gas emissions, pollution reductions, brand development opportunities, and energy price stability, to name a few.

Voluntary markets help develop nationwide renewable energy capacity that exceeds what mandatory markets contribute alone. Ensuring that voluntary markets are separate from and in addition to mandatory markets helps reduce the environmental impact of electricity generation. EPA's Green Power Partnership recognizes organizations that voluntarily buy green power products.

PURCHASING GREEN POWER

Benefits of Purchasing Green Power

An organization's purchased electricity use can be a significant source of air pollution and greenhouse gas emissions. Buying green power can help reduce your organization's environmental impact while also providing the following benefits:

1. Reducing carbon dioxide (CO_2) emissions
2. Reducing some types of air pollution
3. Potentially reducing future electricity price increases
4. Serving as a brand differentiator
5. Generating customer, investor, and/or stakeholder loyalty
6. Creating positive publicity for an organization
7. Demonstrating civic leadership

Variances in Green Power Cost

The price paid for green power can vary widely by the following:

1. Resource type (e.g., solar, wind, biomass)
2. Resource geography
3. Product type (Renewable Energy Certificate [REC]), variable pricing, fixed pricing)
4. Contract duration
5. Vendor

EPA can support an organization in communicating the benefits of green power purchases to stakeholders. EPA can lend valuable credibility to an organization's actions, helping verify that the purchase meets nationally accepted standards for size, content, and resource base.

THE GREEN POWER PURCHASING PROCESS

The following must be included in any green power purchasing process:

Identify a Doer

A doer can play an important role in buying green power. Key doers can be the CEO, or someone else who has the ability to take action. This step is especially critical for larger organizations.

Gather Energy Data

Estimate an organization's purchased electricity quantities by using recent electricity bills or by using a kWh per-square-foot estimate.

Determine the Purchase Scope

Some organizations find that buying green power for a facility, or group of facilities, is the most logical place to start. The purchase scope (facility vs. organization-wide) might influence the type of product you choose.

Evaluate Product Options

There are three eligible product options to choose from when meeting EPA's purchase requirements. In some cases, not all products will be available in a given area.

1. Renewable energy certificates
2. Utility products
3. On-site generation systems

Develop Purchase Criteria

It is helpful to develop a list of criteria when evaluating provider and product options. Criteria can be based on such factors as budget, resource geography, resource base, carbon benefit, or contract type and length.

Solicit Product Providers

Contact multiple providers to determine the going market rate for green power products. Many providers of green power offer other services outside of their primary green power offering.

Develop a Purchase Plan

By developing a purchase plan, it will be easier to assess the value of buying green power together with potential issues.

Buy Green Power

When buying green power, ensure that the contract allows the customer to make environmental claims. For the purpose of partnering with EPA, the purchase must meet EPA's *Partnership Requirements*.

Work with EPA to Capture the Benefits of the Purchase

As a partner, an organization can receive EPA support in making purchase announcements. Partners can receive press release assistance, environmental equivalency statements, and use *EPA's Green Power Partner Mark* on websites and communications materials.

TYPES OF PRODUCTS AVAILABLE

Several different types of green power products are available. The main distinction between product options depends on where the power generation equipment is located—on the power grid or on-site at the facility. An organization can utilize any combination of green power products to meet its goals.

Renewable Energy Certificates (RECs)

Renewable Energy Certificates (RECs) also know as green tags, green energy certificates, or tradable renewable certificates, represent the technology and environmental attributes of electricity generated from renewable resources.

Benefits

- Provides flexibility when green power products are not otherwise locally available

- Lets the customer maintain existing relationships with their utility service providers
- Serves as a practical product option for leased space environments when an organization may not have direct control of its utility service relationship
- Allows the specifying of product sourcing requirements such as resource type and location. Allows scaling the size of an organization's green power purchase
- Price premiums can be less than green pricing products

Additional Considerations

- Does not provide a financial hedge against rising energy costs
- Can be a challenge to communicate the concept of RECs to stakeholders

Utility Products (Green Pricing or Green Marketing)

Depending on whether the facility is in a regulated or competitive electricity market, it may be possible to buy a green pricing product or green marketing product from the electricity provider.

In competitive markets, customers can choose to purchase green marketing products from providers other than the local utility.

In regulated markets, customers may be able to buy a green pricing product from the local utility. In either case, buyers pay a small premium in exchange for electricity generated from green power resources. The premium covers the increased costs incurred by the power provider (i.e., electric utility) when adding green power to its power generation mix.

Benefits

- Consolidates green power purchase and electric service into a single utility bill
- Can offer savings through long-term purchase contracts with the utility service provider
- Is typically sourced from local renewable resources within the utility's service territory

Additional Considerations

- Can be limited in terms of resource base, location, and emissions reduction value
- Can include fuel charges associated with fossil fuel generation

On-site Generation Systems

On-site generation produces electricity from renewable resources using a system or device located at the site where the power is used. Eligible options for on-site generation traditionally include solar photovoltaic (PV), wind turbines, and biomass combustion.

Facilities located near landfills or sewage treatment plants can also explore the option of recovering methane gas for on-site electricity generation.

Benefits

- Provides a predictable fixed price for electricity over the life of the system
- Supplies a visible and tangible environmental commitment to stakeholders
- Is easily communicated to and understood by stakeholders
- Generates clean, renewable electricity at point-of-use
- Can be sized to meet your objectives
- Supports local economies and job creation

Additional Considerations

- Can be limited to states where financial incentives and high energy costs coexist, in order to achieve a financial return on the system within a specific time frame
- Can require a high up-front investment
- Requires maintenance over the life of the system
- Involves more up-front planning and project management resources
- System owners must retain RECs associated with the system in order to make environmental claims.

CERTIFIED/VERIFIED PRODUCTS

Why Certification?

Certification answers the question "Does this product meet acceptable standards for quality?" *Certified* products meet widely accepted consumer and environmental standards and ensure the product purchased comes from eligible renewable resources and meets product-marketing standards.

Certification ensures the quality of a green power product, but also validates the product's environmental attributes. Certification includes standards of conduct for ethical behavior, including marketing claims by suppliers, and requires regular reporting to monitor these claims.

What is Verification

Verification answers the question "How do I know I'm getting what I pay for?" Third-party certification usually carries a requirement for independent verification to document that the amount of green power generated equals the amount of green power sold to customers. Third-party independent auditors apply the verification process to retail and wholesale electricity providers.

The audit verifies that the green power behind the product was produced and placed on the utility grid and helps verify the product's environmental benefit. Verification serves as a form of buyer protection against deception or fraud.

Why Should Certified and Verified Products be Purchased?

Green power products, certified by an independent third party, offer consumers a high level of certainty that they are getting what they pay for. In meeting specific environmental and customer protection guidelines adopted by the certifying organization, an organization can be sure that its purchase meets nationally accepted standards for resource and product quality.

Who Certifies Green Power Products?

Currently, two organizations certify green power products:

1. Center for Resource Solutions' Green-e Program
2. Environmental Resources Trust, Inc.

These organizations have programs in place that not only certify green power products, but also independently verify the products on an annual basis.

Partnership Perspective

While EPA currently does not require green power partners to purchase certified products, they strongly encourage organizations to purchase green power products that are certified by an independent theirparty as a matter of best practice. Certification helps ensure the quality of green power products, and also helps build consumer confidence in the marketplace.

SYNOPSIS—GREEN POWER

Green power and sustainability concepts and processes are here to stay. A company mayor may not believe in or embrace the basic concept. However, it is prudent to know what the processes are and how they will affect overall costs.

At this time, green power and sustainability, at the utility customer level, are primarily voluntary but that will probably change in the future.

Section II

Natural Gas

Chapter 10

Natural Gas, An Overview

NATURAL GAS—THE PURCHASING PROCESS

Natural gas purchasing generally occurs in one of two scenarios. (1) The natural gas customer receives the monthly billing from the natural gas utility and pays it. (2) The natural gas customer orders the natural gas needed on a monthly basis through the services of a provider (broker, marketer, or producer), and arranges for transportation of natural gas on both an interstate as well as intrastate basis.

Neither of these procedures is wrong as long as the procedure results in the least costly natural gas for the customer. The problem is that not many natural gas customers are really certain that the purchasing scenario utilized is the best for obtaining least expensive natural gas.

Natural gas purchasing, for all practical purposes, is generally deregulated on a federal or interstate basis. On an intrastate basis, there are many conditions that exist that, for the most part, translate into higher costs for the commercial or industrial customer.

From a natural gas customer's viewpoint, it is important to sort out all of the potential pitfalls and opportunities that currently, as well as in the future, affect natural gas costs.

Miscellaneous Information about Natural Gas

Question: Natural gas customers are seeing large price variations. Why is this happening?

Answer:

1) Natural gas is increasingly popular for use in homes, schools, businesses, factories, and electric power generation. However, natural gas production has not kept pace with rising demand, and the market price of natural gas reflects the extremely tight balance between natural gas supply and demand.

2) The wellhead price of natural gas was relatively stable during the 1990s—around $2 MMBtu—because supplies of natural gas were plentiful.

   However, recently natural gas prices have varied between $6 and $13 due to weather changes (a record cold winter followed by a mild winter), increased use of natural gas to generate electricity, and public policies make it increasingly difficult for energy producers to keep up with consumer demand.

3) Wholesale natural gas prices remain quite sensitive to weather, both in terms of natural gas supply, such as a hurricane that may affect production in the Gulf of Mexico; and, natural gas demand, such as cold weather that increases usage for home heating.

Question: What steps are natural gas utilities taking to manage and/or reduce price variability?

Answer:

1) Utilize underground storage. Many natural gas utilities purchase natural gas during warm weather months, and store it underground for use on cold winter days. About 20% of the natural gas consumed during the winter comes from storage. This helps ensure enough natural gas to meet the customers' needs.

2) Utilize hedging strategies. More than half of the states allow utilities to use financial tools such as futures contracts and weather risk insurance to stabilize natural gas prices.

3) Contract terms. Natural gas supply managers obtain gas supplies from a variety of sources and under different contract terms to help minimize costs.

Question: What is the real answer to reducing natural gas costs?

Answer:

1) INCREASE SUPPLY—it is in consumers' best interest to do so. Supply and demand are now in a very tight balance, and changes in the weather or economic activity have an almost immediate impact on the wholesale price of natural gas. **More supply = lower prices!**

Question: How do natural gas price variations affect the United States' economy?

Answer:

1) Natural gas meets one-fourth of the United States' total energy needs. Volatile natural gas prices put America at a competitive disadvantage. Natural gas is the backbone of American manufacturing—used to make steel, glass, chemicals; dry paper; process food; and many other products.

Question: Why are natural gas producers not able to keep up with demand?

Answer:

1) Many wells that have produced abundant natural gas for years are becoming depleted.

2) It is sometimes difficult and more costly to extract natural gas from mature reserves. It is important for producers to be able to move to fresh supply areas, and use the best technologies to find and produce more natural gas.

3) Even when producers hold valid leases, they often face months of delays and red tape when getting federal or state permits to start working on bringing energy supplies to consumers.

Deregulation in the Natural Gas Industry

Until recently, the structure of the natural gas industry was rather simple. Natural gas producers explored and produced natural gas, and in general, sold that gas to pipeline companies. The pipeline companies transported the gas across the country and, for the most part, sold it to local natural gas utilities. The utilities then sold the natural gas to their customers.

The federal government regulated the price for which producers could sell their gas to interstate pipelines and also regulated the price for which pipelines could sell the natural gas to utilities. In turn, state and local governmental agencies regulated the price that local natural gas utilities could charge their customers.

Today, there is no federal price control or regulation on natural gas.

The introduction of deregulation, has made the natural gas industry more responsive to customer needs and market forces. Efficiencies, prompted by deregulation can reduce costs for natural gas service for all types of customers.

Competitive Options for Natural Gas Customers

A growing number of natural gas customers have the opportunity to purchase natural gas service on an deregulated basis. Most large-volume natural gas customers—such as electric power generation facilities and manufacturing plants—have found the transition to a more competitive environment relatively straightforward. Almost all industrial customers now have the option to purchase natural gas from a third-party supplier.

AREAS OF NATURAL GAS REGULATION

Natural gas is regulated not only with relation to pricing but also in many other areas as shown below.

Production and Supply

State

- Conservation Laws
    1) Well spacing
    2) Mineral rights
    3) Environmental impact 4) Production rates

Federal

- Environmental Protection Agency (EPA)
    1) Emissions
- U.S. Department of Energy
    1) Authorizes imports and exports
    2) Strategic petroleum reserve
- Bureau of Land Management, Minerals Management Service
    1) Leasing

Natural Gas Pipeline Companies

State

- Regulate intrastate pipeline companies

Federal

- Federal Energy Regulatory Commission (FERC)
    1) FERC regulates interstate pipeline rates
    2) Pipeline construction
    3) Certain pipeline operations

Local Jurisdictions

- Regulate municipal LDC rates and operations

CONTROLLING & REDUCING NATURAL GAS COSTS

Hedging Strategies

Hedging is a strategy to reduce the risk of adverse price movements of natural gas in future periods of time and can be utilized to reduce natural gas price volatility.

One means of doing this is to utilize the New York Mercantile Exchange (NYMEX) futures pricing. By utilizing NYMEX monthly pricing, a long-term (up to 36-month period) fixed pricing structure can be developed. This strategy does not guarantee the absolute lowest cost, but does allow a customer to know their natural gas costs for a future period of time.

NYMEX Details

- Price quotations are for delivery at the Henry Hub receipt point
- Contracts are for 10,000 MMBtu for delivery up to 36 months in the future

NATURAL GAS DEREGULATION

Natural gas deregulation has progressed on a non-uniform basis due to individual state attitudes toward the process. Some states allow deregulation for all customer classes—residential, commercial, and industrial. Some states allow deregulation only for large industrial customers. Some states do not allow any customers to participate in the deregulation process. There is no rationale concerning a particular state's position on the natural gas deregulation process.

Following are potential reasons for state differences in their attitudes toward the natural gas deregulation process:

- States act independently of each other
- Political and economic objectives differ
- Regulatory structures differ
- Market size diversity that will attract energy marketers

NATURAL GAS MEASUREMENT

Natural gas is measured by its heat (Btu) content. It is delivered in volumetric units (Ccf, Mcf), but it is measured and priced according to its Btu value. Natural gas is compared to and is in competition with other fuel sources—electricity, fuel oil, propane, etc.

Different types of energy are measured by different physical units: barrels or gallons for petroleum, cubic feet for natural gas, tons for coal, kilowatt-hours for electricity. To compare different fuels, we need to convert the measurements to the same units. In the United States, the unit of measure most commonly used for comparing fuels is the British Thermal Unit (Btu), which is the amount of energy required to raise the temperature of (1) pound of water, (1) degree Fahrenheit.

It would be difficult, if not impossible to compare (1) cubic foot of natural gas to (1) kWh (kilowatt-hour) of electricity, or (1) gallon of fuel oil or (1) gallon of propane to determine the relative cost of one to the other. The only common value is heat content—Btu. To evaluate the relative cost of natural gas to other fuels, the Btu content of each is compared and priced to determine the true cost of each on a common measurement basis.

Figure 10-1 shows the Btu values of natural gas and several of its potential replacements. By utilizing common Btu costs for each of the various fuels, a true cost for each can be calculated.

SUMMARY

Electricity at $0.087/kWh would seem much less expensive than natural gas at $9.75/Mcf. But when the Btu values of each are compared, natural gas is much less expensive—$9.466/MMBtu versus electricity—$25.49/MMBtu.

Cost is only one consideration of many items that have to be made

Figure 10-1. Heat (Btu) Value of Fuels

Btu Value of Common Fuels

FUEL TYPE	*AMOUNT*		*TYPICAL Btu Value*
Natural gas	1 cubic foot	=	1,030 Btu
Electricity (1,000 Watts/hr)	1 kilowatt-hour	=	3,412 Btu
Propane	1 gallon	=	91,000 Btu
Gasoline	1 gallon	=	124,000 Btu
Heating oil	1 gallon	=	139,000 Btu
Diesel fuel	1 gallon	=	139,000 Btu
Crude oil (42 gallons)	1 barrel	=	5,800,000 Btu
Fuel oil (42 gallons)	1 barrel	=	6,287,000 Btu
Coal (2,000 lb)	1 short ton	=	20,681,000 Btu

Figure 10-2. Cost Comparison Example

1. 1,000 cu ft (Mcf) of natural gas costs $9.75 at customer's meter point.
2. (1) kWh of electricity costs $0.087 at customer's meter point
3. Convert to cost per 1,000,000 Btu (MMBtu)

NATURAL GAS

$9.75/Mcf + 1,030,000 x 1,000,000 x $9.75 = $9.466/MMBtu

ELECTRICITY

$0.087/kWh (1,000,000 + 3412 x $0.087) = $25.49/MMBtu

when determining what fuel to utilize for a specific purpose. Since fuel is a continuing cost and will escalate over time, it is important to understand its impact on overall costs.

SYNOPSIS—NATURAL GAS—AN OVERVIEW

1. The purchasing process
2. Miscellaneous information

3. Regulation
4. Competitive options
5. Deregulation
6. Hedging
7. Measurement criteria

This information presented in this chapter is important to the natural gas cost reduction process since it is basic to understanding the overall natural gas industry.

This chapter, together with the chapters that follow are similar to the assembling of instructions—*not necessarily exciting but critical to the finished product*. Utilizing this information will result in a properly working process, which is a cost-effective natural gas strategy.

Chapter 11

Natural Gas Regulation

NATURAL GAS—THE PLENTIFUL FUEL

Natural gas is one of the most plentiful fuels that is found in the contiguous United States. It is surpassed only by coal in domestic reserves.

Depending upon which natural gas expert you want to listen to, it appears that, based upon current usages in the United States' reserves, there is between 50-100 years supply of known natural gas reserves. In addition to the United States reserves, there are considerable additional natural gas reserves in two other bordering countries—Canada and Mexico.

NATURAL GAS REGULATION

Regulation in General

Generally speaking, utilities are considered to be regulated monopolies. A *regulated monopoly* is an entity that has a protected service territory and has virtually no competition since no other provider of the same commodity can compete for the customer of the protected service territory.

Since a position of no competition can lead to abuse in both cost and service, a check and balance system is in place that requires utilities to be held accountable for costs, services, etc. Utilities are regulated in at least two areas, *interstate* (between states) and *intrastate* (within a state).

Interstate Regulation (Between States)

Natural gas interstate regulation is the responsibility of the Federal Energy Regulatory Commission (FERC). This agency was created in 1977 and has the responsibility for the oversight and regulation of interstate transportation policies and rates concerning electricity. Since this agency has these responsibilities, it would be well to understand its impact on rates and transportation conditions.

FEDERAL ENTITIES

Federal Energy Regulatory Commission (FERC)

Regulation—Federal Regulation

This type of regulation is of federal origin and generally does not include intrastate regulation. Since LDCs are intrastate in operation, they are not directly within FERC's jurisdiction. The federal government has the overall responsibility of regulation and operating procedures only on an interstate basis.

As a rule, interstate entities wholesale the majority of their natural gas to for-profit investor-owned companies, municipalities, and cooperatives who in turn are regulated by their respective regulatory agencies. When direct sales are made to customers, regulation parameters are by federal government guidelines.

About FERC

Mission

The Federal Energy Regulatory Commission regulates and oversees energy industries in the economic and environmental interest of the American public.

Goal

FERC's goal is to provide dependable and affordable energy through sustained competitive markets.

How FERC Operates

The Federal Energy Regulatory Commission (FERC) is an independent agency that regulates the interstate transmission of natural gas. As part of that responsibility, FERC:

1. Regulates the transmission and sales of natural gas for resale in interstate commerce.
2. Approves the siting of and abandonment of interstate natural gas facilities, including pipelines, storage, and liquefied natural gas.
3. Oversees environmental matters related to natural gas, and hydroelectricity projects and major electricity policy initiatives.
4. Administers accounting and financial reporting regulations and conduct of regulated companies.

What FERC Does Not Do

Areas considered *outside* of FERC's jurisdictional responsibility include:

1. The regulation of retail natural gas sales to consumers.
2. The regulation of activities of municipal and cooperative power systems, and federal power agencies (See Figure 11-1).
3. The responsibility for pipeline safety.
4. The regulation of local distribution pipelines of natural gas.
5. The development and operation of natural gas vehicles.

FERC's Oversight of Natural Gas Pipelines Includes

1. Regulation of pipeline, storage, and liquefied natural gas facility construction.
2. Regulation of natural gas transportation in interstate commerce.
3. Issuance of certificates of public convenience and necessity to prospective companies providing energy services, or constructing and operating interstate pipelines and storage facilities.
4. Establishment of rates for services.
5. Regulation of the transportation of natural gas as authorized by the NGPA (Natural Gas Policy Act).
6. Oversight of the construction and operation of pipeline facilities at U.S. points of entry for the import or export of natural gas.

Natural Gas Areas Regulated by FERC

1. Establishes and enforces rates and charges for natural gas transmission and sales for resale (very important to natural gas deregulation)
2. Establishes and enforces operational characteristics, charges for natural gas interconnections rates, and
3. Establishes and enforces rates and charges for distribution and sale of natural gas

Intrastate Regulation (within the state)

Natural gas intrastate regulation occurs within the borders of a state and is the responsibility of one of at least three types of regulatory bodies. The type of natural gas utility that serves a customer determines who regulates what.

Figure 11-1. Federal Power Agencies

1. Alaska Power Administration
   Juneau, AK
   www.energy.gov

2. Bonneville Power Administration
   Portland, OR
   www.bpa.gov

3. International Boundary & Water Commission (U.S. & Mexico)
   El Paso, TX
   www.ibwc.state.gov

4. Southeastern Power Administration
   Elberton, GA
   www.sepa.doe.gov

5. Southwestern Power Administration
   Tulsa, OK
   www.swpa.gov

6. Tennessee Valley Authority
   Knoxville, TN
   www.tva.gov

7. U.S. Army Corps of Engineers
   Washington, DC
   www.usace.army.mil

8. U.S. Bureau of Indian Affairs
   Mission Valley Power
   Polson, MT
   www.missionvalleypower.org

9. U.S. Bureau of Indian Affairs
   San Carlos Irrigation Project
   Coolidga, AZ
   www.energy.gov

10. U.S. Bureau of Reclamation
    Washington, DC
    www.usbr.gov

11. Western Area Power Administration
    Golden, CO
    www.wapa.gov

For-profit

For-profit natural gas utilities are regulated on a statewide basis by a commission or group of individuals appointed or elected at the state level. This commission, or group of individuals, regulates all for-profit natural gas utilities, regardless of their geographic location within a given state.

Municipal

Municipal utilities (utilities owned and operated by a city or county) generally are self-regulated. Normally, municipal utilities are autonomous and structure their rates as they see necessary with minimal or no state supervision or legislation.

Cooperative

Cooperative utilities are similar to municipal utilities in that they are generally self-regulated with little state oversight. This type of utility often comes into being when a for-profit utility does not provide service to a given area, generally because of a lack of a sufficient profit making opportunity.

A cooperative utility is usually created when a group of potential users form their own utility to provide a commodity (electricity, natural gas or water/sewer) for themselves that otherwise would not be available.

Similarities

Both municipal and cooperative utilities are generally self-governing with little state intervention except with relation to rate case procedures and public notification guidelines. Ordinarily the state will require that affected customers (intervener groups) follow normal rate case protocol with relation to due process and input.

It is very important to know what is happening at the state regulatory level because of the impact of rate case decisions on customer costs.

At the federal (FERC) (interstate) level, natural gas is more deregulated than at the state (intrastate) level. Regulation, at the intrastate level, is generally more restrictive and, in many instances, results in rate provisions that cause the customer rate to be more costly than it should be. An awareness of pending cases is important so that a strategy can be developed based upon fact.

Areas Regulated by Intrastate

Intrastate Rate Structures and Pricing Criteria

Types of service that are available in most the local distribution company (LDC) service areas are as follows:

Firm—A firm rate is not interruptible except for force majeure circumstances.

Interruptible—An interruptible rate can be interrupted for non-Force Majeure circumstances—cost of gas, load on system, etc.

Negotiable—A negotiable rate competes with alternate sources of natural gas.

Firm Distribution—A firm distribution rate is the transportation of the retail customer's natural gas through the serving utility's distribution system on a firm basis.

Interruptible Distribution—An interruptible distribution rate is the transportation of the retail customer's natural gas through the serving utility's distribution system on an interruptible basis.

Three Types of LDC Regulation on an Intrastate Basis

For-profit (investor-owned companies)

Regulation—All for-profit investor-owned LDCs that operate within state boundaries

The for-profit utility regulation is on a statewide basis since intrastate regulation of LDCs concerns the activities of an LDC within that particular state. State agencies regulate the intrastate transportation and operation of the natural gas LDC. Since natural gas may be distributed from a location outside of the boundaries of the state in which the customer is situated, there are both federal and state regulations that apply.

The state regulatory agencies usually take the form of public service commissions (PSC) or public utility commissions (PUC). The functions of these entities are to regulate the intrastate distribution and operation of LDCs. These agencies also determine and approve individual LDC rates of return, grant franchises to LDCs for specific areas of operation, and in general, regulate the operation of the LDCs within the individual state.

Although PSC or PUC structures are the most common forms of state regulation, other methods are used. In some states, these commissions regulate LDCs only outside the incorporation limits of a municipality or city such as in the states of Georgia, Texas, etc.

Also, there are some strange situations that occur in a few states. For instance, in the State of Texas, the Public Utility Commission regulates electricity outside of municipalities. However, outside of these same Texas municipalities is an entity called the Railroad Commission of Texas, which regulates the natural gas utilities.

Overview—For-profit (investor-owned) LDC Regulation

1. *Intrastate regulatory body.* State regulatory agency.
2. *LDCs in the classification.* Any LDC that is in business for the stated purpose of making a profit and is owned by investors through the purchase shares of stock.
3. Regulation process. Any rate change requires at least the following items:
   a) A public notification of intent to change a rate.
   b) Adequate notification period prior to actual rate case presentation (this period defined by state law) to allow interested parties to study the merits of the change request.
   c) Presentation of rate requests at a hearing open to the public before the appropriate regulatory agency.
   d) Allow input from interested customer groups (interveners) relating to rate change requests. The state regulatory body, based upon testimony presented at rate case hearings, approves actual rates that will be put into effect.

Municipalities

Regulation—Generally, self-regulated

When municipalities regulate LDCs, they are as a rule self-governing—that is, municipalities are not subject to state regulatory rulings.

Generally, when municipalities undertake the duties of the LDC, they are purchasing the natural gas on a wholesale basis and then retailing the natural gas to their residential, commercial, and industrial customers. This scenario is true with natural gas, but in the case of water and sewage, the municipality usually has control or jurisdiction over the entire process.

Municipal LDCs are generally presided over by an LDC commission or a board of appointed and/or elected members. As in the case with all LDC regulatory agencies, due process must occur before changes can be made with respect to rates and conditions under which natural gas is provided.

This means that a public notice must be given and adequate time allowed for public input prior to a change being instituted. Typically, municipal LDCs do not have as many rate classes or options as do for-profit investor-owned LDCs since municipalities generally do not have as diverse a class of customers.

Overview—Municipal LDC Regulation

1. *Regulatory body*. Generally self-regulated by the body or agency selected by the municipality to oversee LDC matters. (Can be separate for each utility regulated.)
2. *LDCs in this classification*. Any LDC operated and regulated by a municipality on a not-for-profit basis.
3. *Regulation process*. Similar to for-profit investor-owned LDCs.
   a) Public notification
   b) Adequate notification period
   c) Presentation of rate request
   d) Allow input from interested customer groups

Rural Cooperative LDCs

Regulation—Generally, self-regulated

Cooperative LDCs are formed generally when a for-profit investor-owned LDC elects not to serve a geographic area or customer base. Cooperative LDCs usually serve rural areas where there is not a large customer load base. Generally, power is purchased at wholesale from a for-profit LDC and distributed by the cooperative's pipes to the individual customer's location.

Cooperative LDCs are like municipal LDCs in that they are self-regulated but also are required to provide due process before instituting changes in the LDC rate base. They are also very different from any other type of LDC since they are classified as a cooperative entity.

The term cooperative, as far as LDCs are concerned, literally means that each customer is a part owner of the LDC, and as such, at least in theory, has their proportionate say in how the LDC is operated.

Also, cooperatives are similar to all other types of LDCs in rate change cases. Their procedure is to propose rate changes, hear customer input, and all other due process practices before actually instituting rate changes. Cooperative LDCs are smaller in terms of total energy supplied than are municipal LDCs, and are generally located in rural types of service areas.

Overview—Rural Cooperative LDC Regulation

1. *Intrastate regulatory body*. Generally self-regulated. All customers are part owners of the LDC and as such have their proportionate say or vote based upon their usage in relation to other customers. In practice many times, a board of overseers is appointed to represent cus-

tomer interests with respect to the LDC operation.

2. *LDCs in this classification*. Any LDC defined as cooperative and that is owned and operated by the customer served.
3. *Regulation process*. Similar to other types of LDCs and includes:
   a) Public notification
   b) Adequate notification
   c) Presentation of rate request
   d) Allow input from customers

THE INFLUENCE INTRASTATE REGULATION HAS ON NATURAL GAS COSTS

State agencies regulate the intrastate distribution of natural gas. A LDC typically uses a combination of INTER (between states) and INTRA (within a state) components. Consequently, both federal and state regulation occurs.

Since state agencies provide predominate day-to-day natural gas regulatory functions, it is important to know how these regulatory agencies operate with relation to the customer. One of the factors in intrastate regulation is the number of LDCs that do not offer actual cost-of-service rate schedules. It is of no value for a customer to have interstate access to less expensive natural gas if the intrastate LDC does not allow transportation of that natural gas through their individual systems.

Regulation of Natural Gas

See Figure 11-2.

Natural Gas Processing

See Figure 11-3.

SYNOPSIS—NATURAL GAS REGULATION

1. Natural gas regulation
   a) Interstate
   b) Intrastate
2. The Federal Energy Regulatory Commission (FERC)
3. Federal power agencies

4. For-profit LDC regulation
   a) State intervener groups
5. Municipal LDC regulation
6. Rural cooperative LDC regulation
7. Overview of LDC regulation
8. Synopsis of natural gas regulation

The information in Chapter 11 is basic to the efficient, cost-effective natural gas purchasing process. Do not disregard this basic data, as it is fundamental to a successful natural gas cost reduction strategy.

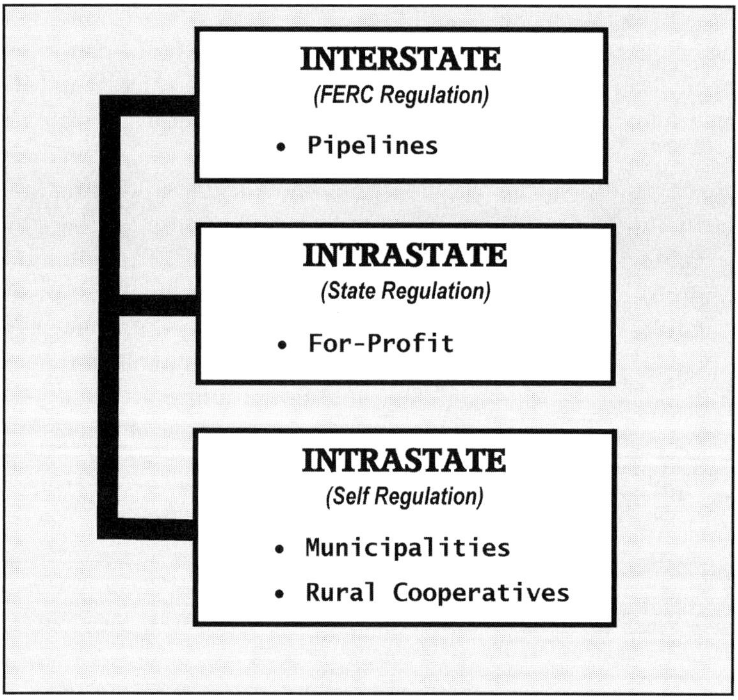

Figure 11-2. Local Distribution Companies (For-profit, municipal, rural cooperative)

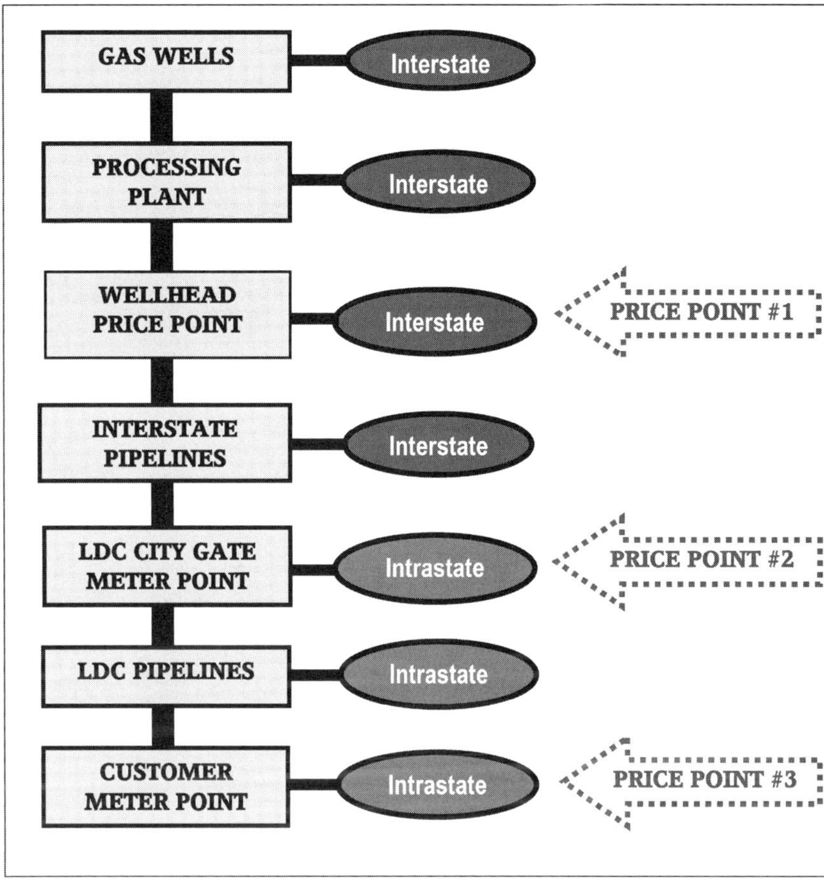

Figure 11-3. Flow of Natural Gas

Chapter 12

Reducing Natural Gas Costs

PRODUCING A COST REDUCTION STRATEGY

The first step in this undertaking is the obtaining of accurate information relating to the subject being investigated. In the case of natural gas, the first step in understanding begins with the obtaining of data that relates to the pricing on a unit basis of the commodity (natural gas) being purchased.

Since all LDCs are regulated in one way or another, rate schedules and pricing data, as it relates to customers, must be available to any interested party.

Items Needed from the LDC and State Regulatory Agency

Listed following are the mandatory basic informational items that must be obtained before any understanding of natural gas rates will be realized. All of these items are a matter of public record and must be made available to anyone who requests them. They are typically available from websites, the LDC, and regulatory agencies

Generally, the most logical place to obtain this information is from the LDC or the LDC regulatory agency. If a request is received from a customer of the LDC, there generally is no problem or cost involved. But if there should be a charge, then it should be reasonable and reflect the time and material required furnishing the information. However, if a request is received from a non-customer, some problems may arise both with respect to availability as well as the potential for a cost being assessed for the material.

All LDC rate information that is approved by a regulatory body for use in determining rates and conditions to which a customer is subject, must be a matter of public record, and as such, available for public inspection.

Typically, the LDC service representative responsible for the customer involved is contacted and a request is made for the information needed, and many times this information can be e-mailed to the customer. The importance of obtaining the following information cannot be overstated since these items are basic to understanding LDC costs.

With relation to information that is required from the state, the best way to proceed is to contact the state agency involved directly. As a rule, there should be no problem in obtaining the sales tax or the economic development and/or enterprise zone information from the state.

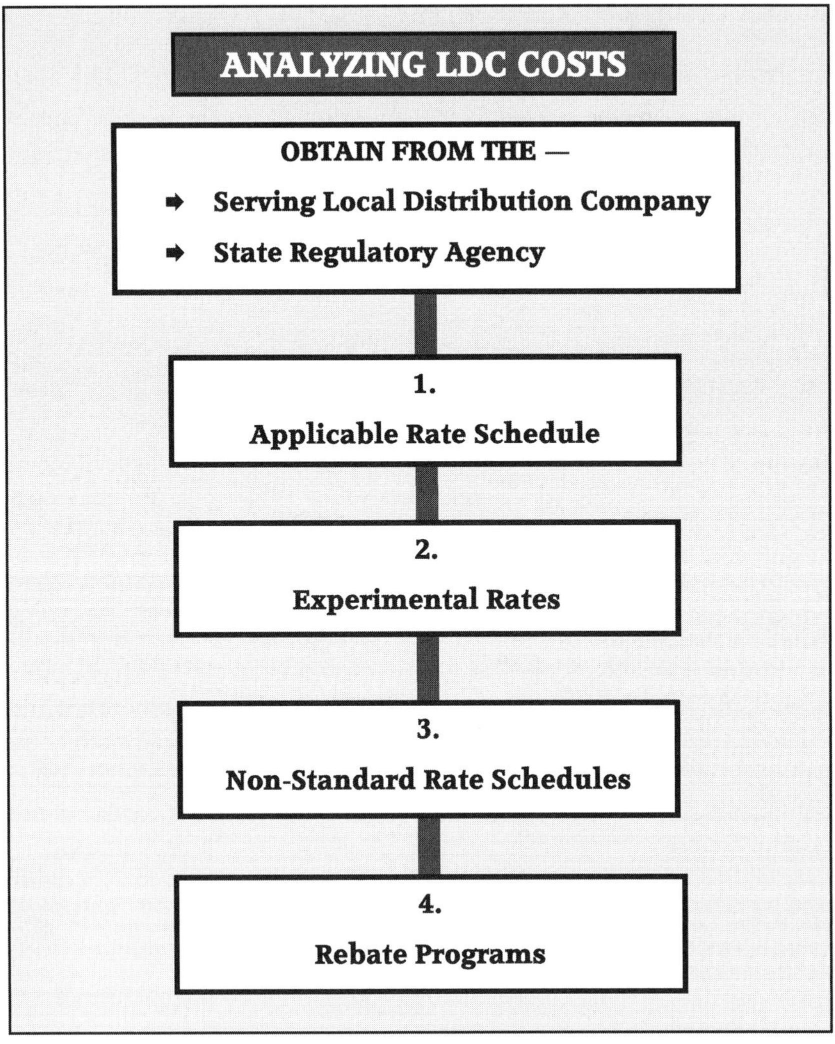

Figure 12-1. Basic LDC Data Needed

BASIC LDC DATA NEEDED OBTAINED FROM THE SERVING LDC

Applicable Rate Schedule

A complete rate schedule covers all rates, terms, and conditions that are approved in rate cases, The classes of customers addressed are residential, commercial, and industrial. These schedules can range from a few sheets to several hundred sheets in length. Contained in this information will be all data relating to customer rates, costs, terms for service, etc.

The importance of this source document cannot be overemphasized since it is mandatory for an understanding of natural gas costs, Make certain that a request is made for a *complete* rate schedule since LDCs tend to provide only the particular rate that currently applies to the customer making the request. Having available a *complete* rate schedule cannot be over emphasized since only then can comparisons be made between different rates and options. A typical complete rate schedule will contain the following items:

1. Complete list and explanation of all customer rates available
2. Complete list of all items or riders that modify or change rate costs
3. Alternative rates that may be available on a *customer request* basis for certain customer classes
4. Information on special rates that may be available in certain circumstances
5. Complete explanation as to how all cost components of natural gas usage are measured and applied. Complete rate schedules remain in effect until a new rate case is filed and approved by the appropriate regulatory agency. Only one complete schedule is required since all customer classes are addressed therein.

Analysis of Rate Schedule—NGSSF

1. *Rate Schedule.* This rate schedule (NGSSF) identifies the rate schedule that this LDC assigns to their regular natural gas sales service customers who have no backup availability. This is a typical rate schedule utilized where the serving LDC provides both services and commodity (natural gas) to a customer on a firm, non-interruptible basis.

2. *Applicability.* This rate is applicable to any LDC customer having the required usage characteristics that requests firm commodity and transportation services from the serving LDC.

Figure 12-2. Example—Natural Gas Rate—Firm Service

1. RATE SCHEDULE: NATURAL GAS SALES SERVICE—FIRM (NGSSF)

2. APPLICABILITY: This service is offered for any customer who executes an agreement to have this LDC be responsible for supplying all the customer's natural gas requirements, and who plans to use this service at all times for operating any of its equipment designed to use natural gas. This service is not intended for customers desiring any flexibility in choice of fuels or natural gas suppliers.

3. AVAILABILITY: This service is available to any customer who is either:
   A. A new customer who executes an agreement for an initial term of up to two (2) consecutive years through October 31 of the second fiscal year; or
   B. An existing customer of the company who executes an agreement at least eight (8) full calendar months prior to the November 1 that the customer intends to take service under this service schedule.

Service to a new customer with a maximum demand of over 1.000 therms a day is subject to the company's ability to secure sufficient gas and capacity.

Service hereunder is on a perpetually renewable fiscal year-to-year term commencing November 1 and ending October 31. Should a customer desire to take service under another service schedule and avoid automatic renewal, the customer must enter into a completed agreement under a different service schedule no later than the March 1 (i.e., eight (8) months advance notice) prior to the November 1 on which the new service schedule is scheduled to commence. The customer's rights pursuant to said notice will terminate as of the next January 1 if service under another service schedule is not implemented. Should a customer decide after March 1, and before November 1, that it does not desire to take service under its new service schedule, then the company will provide gas service under the appropriate Default Service Schedule. Nothing in this paragraph shall limit the company's ability to immediately implement changes authorized by the Public Service Commission.

This class schedule is for customers who need natural gas service(s) for an actual annual usage per the table following. Service is on a perpetually renewable fiscal year-to-year term commencing November 1 and ending October 31. At the end of the annual billing cycle nearest May 1 each year, any customer using less the minimum usage or more than the maximum usage per the table below will no longer be eligible for this class schedule after October 31 of that year.

Page 1

Figure 12-2 (*Continued*)

Class Schedule	*Service*	*Annual Minimum Therms*	*Annual Maximum Therms*
NGSSF	**Firm Sales Service**	**50,000**	**No Limit**

4. USAGE: Distribution cost of gas and miscellaneous adjustments:
   A. Usage Cost. The usage costs are calculated by taking the distribution cost plus any cost of gas and take-or-pay adjustments. (See Rate Schedule 176-B to determine effective usage cost.)
   B. Distribution Cost. Minimum and maximum distribution costs under this service schedule. (See Rate Schedule—(191-A)
   C. Cost of Gas Adjustment. Subject to a cost of gas adjustment Rate—(See Rate Schedule 112-A)
   D. Miscellaneous Adjustments. Subject to miscellaneous adjustments—(See Rate Schedule 193-C)

5. MANDATORY USAGE CHARGE: (Penalty) Any customer using alternate fuel or natural gas provided by another supplier for equipment designed to use natural gas when natural gas is available to the customer under this service schedule will be charged $0.0550 per therm for each therm of natural gas replaced by alternate fuel or other natural gas supplies. The Btu values for various alternate fuels, which shall be used to determine the amount of natural gas replaced, are set out in the company's rules and regulations found on Rate Schedules—187-F, 315-J, and 402-B.

6. UNAUTHORIZED USAGE CHARGES. For unauthorized usage during periods of non-availability, depletion, interruption, or curtailment of service available to the customer under this or any other service schedules(s), in addition to the regular scheduled rate for this service, the customer will be charged a penalty as found on Rate Schedule 170-C.

7. CHARACTER OF SERVICE. Character of service provisions will apply as follows:

   A. The character of service provisions are set forth in the customer's applicable class schedule, including the company's rules and regulations as found on Rate Schedules 167-C to 169-D and 260-A to 271-F.

   B. Service under this schedule contemplates continuous month-to-month use. Customers who have their meters turned off and back on within a twelve (12) month period shall pay the Minimum Charge applicable to the customer for the period while service was not being used.

Page 2

Figure 12-2 (*Continued*)

8. LATE PAYMENT CHARGES. The amounts billed are payable on or before the due date stated on the bill. A late payment charge per Rate Schedule 250-B is applied when payments are not received in the Company's office on time.

9. MINIMUM CHARGE. The monthly minimum charge is the sum of any charges due under this rate schedule.

10. SPECIAL TERMS AND PROVISIONS.
    A. The eight (8) month minimum notice requirement may be waived for customers:
        (1) At the company's sole discretion, and
        (2) If there will be no undue detriment to the company's existing LDC Sales Service (USS) customers, and
        (3) If a customer transferring to a non-NGSSF service from another service, agrees to pay any peak and non-peak demand costs for every therm used for the remainder of the current agreement. Such demand costs are based on the PGA filing in effect at the time of the request before any gas cost adjustments.
        (4) If the customer agrees not to return to the previous service selection—(i) for a minimum period of six (6) months or until November 1, whichever is earlier, and (ii) until points (1) and (3) above have again been satisfied.

    B. It is the company's intention to waive the eight (8) month notice and permit a customer to avoid the conditions of (1) above when doing so will result in the company's resulting total capacity being within planned limits. A customer not meeting the conditions in (1) above that wants to go to transportation service earlier than is otherwise permitted will have the ability to place its account in a queue, upon completion of an agreement, no later than August 14.

    C. On August 15 of each year, the company will determine whether or not it has the ability under its supply plan to permit customers in the queue to go to transportation service without a capacity assignment or whether a capacity assignment will be required. Customers in the queue will have a one-time opportunity within fifteen (15) days notice from the company to confirm their intention, based on the condition set by the company to go to transportation service as of the next November 1. A decision to not go to transportation service will result in the customer continuing to receive service under

Page 3

Figure 12-2 (*Continued*)

this service schedule. Failure to comply with the company's Rules and Regulations will result in the customer losing its ability to go to transportation service as of November 1 and will require the customer effective November 1 to take service from the company under the appropriate default service schedule. Customer not wanting to wait for the determination set out in the above paragraph, can either themselves or through a representative, accept a capacity assignment determined solely by the company. The duration of the assignment will be for a minimum of twelve (12) months up to the remainder of the term of the contract being released.

D. If the customer has more than one meter combined for billing purposes utilizing the provisions of (1) or (2) above, the customer will have the following options:
   (1) The customer can install telemetering on all the meters or can make customer side of the meter changes such that all the consumption qualifies for Rate Schedule T 5-2.
   (2) The customer can install telemetering on any meter(s) and have such meter(s) qualify for Rate Schedule T 5-2 with the remaining meter(s) qualifying for the appropriate class and resulting sales service.

Page 4

3. *Applicability*. This rate is available to any customer in either of two categories:
   A. New customer who executes an agreement for an initial term of two years.
   B. An existing customer who executes an agreement for at least 8 months.

Also, listed in this section are the various rules and requirements for this rate. One of the more important items is the annual minimum therm usage requirement—50,000 therms or 5,000 dekatherms. Although it is stated in the applicability section (section #2) that this rate is available to any customer. In actuality, it is only available to a customer who uses at least a minimum of 50,000 therms annually.

4. *Usage, Distribution, Cost of Gas, and Miscellaneous Adjustments.* Here is described the various component of this rate that determines the

total cost of natural gas utilized through this rate schedule. There are four items that are utilized to calculate these costs, but there no costs are actually shown here. To determine what the actual costs are, it is necessary to utilize at least three additional Rate Schedules 176-B, 193-C, and 112-A.

A. <u>Usage Cost</u>. These costs include distribution, cost of gas, and take-or-pay adjustments. The details and specific costs are found in Rate Schedule 176-B.
B. <u>Distribution Cost—Minimum/Maximum</u>. In addition to the rate schedule referenced in Item A, there is also a Rate Schedule 191-A that details the range of distribution costs that are allowed by the appropriate regulatory agency.
C. <u>Cost of Gas Adjustment</u>. This cost is typically a pass-through adjustment by the LDC to compensate them for natural gas price variations either greater or lower than the standard cost they have in the base rate. Adjustments of this type may result in a credit or additional cost on a particular monthly billing. To determine how this cost of gas adjustment is structured, Rate Schedule. 112-A would be utilized.
D. <u>Miscellaneous Adjustments</u>. This area covers items not specifically addressed in this rate schedule, but could apply in certain situations. To determine what these miscellaneous adjustments could be, Rate Schedule 193-C would be utilized.

5. *Mandatory Usage Charge (Penalty).* Here is described what happens if the customer does not utilize natural gas available in this rate schedule. If the customer chooses to utilize some other commodity for equipment that could utilize natural gas available through this rate schedule, the LDC will impose a penalty per unit of natural gas not utilized through this rate schedule.

   This rate schedule is known as full service. It requires the customer to utilize only natural gas provided through the rate schedule. The details of the regulations concerning this provision are found in Rate Schedules 187-F, 316-J, and 402-B.

6. *Unauthorized Usage Charges.* Described and identified here are the periods where continued usage of natural gas will result in penalties. For penalty provisions, it is necessary to utilize Rate Schedule 170-C.

7. *Character of Service.*
   A. Character of Service Provisions. The specific characteristics are not explained but reference to Rate Schedules 167-C, 169-D, 260-A, and 271-F are provided. To determine what the specific character of service provisions are, it will be necessary to read these four referenced rate schedules.
   B. Month-to-month Use. This rate schedule is designed for customers who use natural gas on a continuing month-to-month basis—not on a periodic or non-continuing basis.

8. *Late Payment Charges.* There are late payment charges on delinquent payments. To determine how much the charges are and how they are assessed, you must have Rate Schedule 250-8.

9. *Minimum Charge.* These charges are simply the total of costs that the customer incurs for the billing month involved.

10. *Special Terms and Provisions.* This section addresses four separate special terms and/or conditions that the LDC can extend to customers at either the LDC's sole discretion or if there will be no undue detriment to other existing customers.
   A,B,C Waiver of (8) Month Minimum Notice Itemized in Item #3—Service Availability. The various exclusions are listed and explained in these three sections (A, B, and C).
   D. Multiple Meter Points. Here is described the process the customer can utilize to combine multiple meter points. These provisions can be very important if a customer has multiple individual meter points and no single meter point utilizing a sufficient quantity of natural gas to qualify for this rate schedule as itemized in Item #3 Availability (50,000 therm minimum annually). Two options are available:
      1) Installation of *telemetering to aggregate all meter point usages* into a usage quantity that will meet the 50,000 therm minimum annual usage threshold.
      2) Customer can *telemeter any single meter* point that has sufficient usage to qualify for the 50,000 therm minimum annual threshold. All other meter points can be left as is and be billed under the appropriate rate schedule based upon their individual usage characteristics.

In this particular example (developed from an actual LDC rate schedule), there are many items that need to be considered by a customer. Also, there are many other rate schedules that need to be utilized to completely understand this rate schedule in its entirety.

Item #	*Rate Schedule Needed*
4-A	176-8
4-8	191-A
4-C	112-A
4-D	193-C
5	187-F, 316-J, 402-8
6	170-C
7-A	167-C, 169-D, 260-A, 271-F
8	250-8
10-D	T 5-2

To fully understand this rate schedule, you need to obtain a total of fifteen (15) separate rate schedules—this rate schedule (NGSSF) plus fourteen (14) other referenced rate schedules. It takes time to completely evaluate and understand a rate schedule of this type.

BASIC LDC DATA NEEDED— OBTAINED FROM THE STATE REGULATORY BODY

Experimental Rate Schedules

Experimental rate schedules are not normally contained in complete rate schedules since they are developed on an experimental basis by LDCs, and are not mandated for any particular customer. These types of rates are not available from all LDCs; however if they are, they can be a source of cost reduction potential. Experimental rates are created by the LDC and approved on an *experimental basis* by the applicable regulatory agency.

The experimental category allows the LDC to evaluate the potential for a different type of rate structure. Since these rates are never mandated, they are used only on a customer voluntary basis. If the customer chooses an experimental rate and it results in an increased cost, then the LDC may not assess any charge higher than what would have resulted from the regular schedule of rates.

If an experimental rate proves successful, the typical next step is to include it as an optional rate in the base rate schedule and not mandatory for any customer class. Since the final step is to change the *optional classification* to a *mandatory* category in the base rate schedule, it remains important to keep up-to-date on experimental rates since long-term they have a way of becoming mandatory for some customer classes.

Determine whether experimental rates are available. If they are, obtain a copy and evaluate the short-term, as well as long-term, implications in case these rates are later included in the base rate schedules as *mandatory*.

STEP #1	Experimental Rate
	↓
STEP #2	Optional Rate
	↓
STEP #3	Mandatory Rate

Analysis of Rate Schedule—Exp-1-IT

The Experimental, Interruptible Transportation Rate allows a customer to reduce natural gas costs. However, it has rather serious penalties for variations between the quantity of natural gas nominated (ordered, and the quantity actually used on a daily basis. An analysis of this rate follows:

1. *Rate Schedule*. This rate (EXP-1-IT) designates the rate case identification number assigned by the regulatory agency. If a customer wants to examine data related to this rate case filing, they can obtain it from either the LOG or the regulatory agency.

2. *Applicability*. This section addresses the type of customer that can be served on this rate. In this particular rate, the customer must use more than 4,000 Ccf on a daily basis. Also, the customer must contract for their natural gas supply through a third party (marketer).

Figure 12-3.
Example—Experimental, Interruptible, Transportation Rate (Marketer-Supplied Natural Gas)

1. RATE SCHEDULE: EXP-1-IT (EXPERIMENTAL, INTERRUPTIBLE TRANSPORTATION)

2. APPLICABILITY:
Available to any customer on an experimental basis for a minimum of 1 year whose natural gas requirements are more than 4.000 Ccf daily on an Interruptible basis for the delivery of natural gas *owned by the customer* through a LDC pipeline to a meter location on the customer's premise.

3. RATE:
Monthly Customer Charge: $150.00
Commodity Transportation:

First	4,000	Ccf	$ 0.095/Ccf
Second	10,000	Ccf	$ 0.061/Ccf
All over	20,000	Ccf	$ 0.042/Ccf

4. SPECIAL CONDITIONS:
   1) Customer must have arranged for the purchase of natural gas other than LDC's pipeline supply and for its delivery to an LDC pipeline.
   2) Customer must have and maintain adequate standby facilities, and have available sufficient fuel supply to maintain operations during periods of curtailment.
   3) Customer agrees to curtail the use of natural gas when requested by LDC.

5. BTU ADJUSTMENT:
Customer billed usage in Ccf volumes will be adjusted when Btu content of delivered natural gas varies from 1.000 Btu's per cubic foot.

6. NOMINATION:
By 8:00 a.m. on a daily basis, customer must directly advise LDC's Natural Gas Control Department of the volumes to be delivered on its behalf from the LDC pipeline to the customer's premises during the following natural gas day commencing at 12:00 (midnight).

Page 1

Figure 12-3 (*Continued*)

7. BALANCING:
Customer and customer's natural gas supplier are responsible for balancing nominations and deliveries to the pipeline transporter's system. When daily volumes of natural gas delivered on behalf of customer to LDC's pipeline receipt points or natural gas received at customer's designated delivery point differ, (above or below daily nominated volumes of natural gas) the customer is out-of-balance. Customer will correct out-of-balance occurrences as provided for in Item #8 following.

8. PENALTY PROVISIONS:
   1) For balancing: When daily volumes are out-of-balance by more than ±10%, a penalty of $0.50 per Ccf, adjusted for Btu content, will be assessed. When monthly volumes of natural gas delivered on behalf of customer to LDC's pipeline receipt points differ from customer use at designated delivery points during any billing month, the customer will correct the imbalance during the next billing month. At the end of the next billing month, any remaining imbalance will be assessed a penalty of $0.15 per Ccf, adjusted for Btu content, per day.
   2) Customer may avoid Penalty Provisions for Balancing listed above, and Penalty Charges, when deliveries on pipeline transporter's system to LDC's pipeline(s) differ from customer's nominations, contained in pipeline transporter's rate schedule for Experimental Interruptible Transportation Service by electing LDC's *Optional Transportation Balancing Service.* (Schedule OTBS-142)
   3) For Unauthorized Use of Natural Gas: Penalty for unauthorized use of natural gas during periods of curtailment will be LDC's prevailing Energy Charge, including costs of purchased natural gas, plus $1.00 per Ccf for all natural gas used. Further LDC shall have the right to shut off supply of natural gas at customer meter point in the event of failure to curtail use after being notified.

9. DUE DATE:
Bill payment is due fifteen (15) days after the bill processing date.

10. LATE PAYMENT CHARGE:
Delinquent amounts are subject to a late payment charge of 2% or $1.00, whichever is greater. No late payment charge will be applied if the delinquent amount is $10.00 or less.

11. UNIT OF MEASUREMENT:
The unit of measurement shall be a cubic foot of natural gas at an absolute pressure of 14.73 psia, and a temperature of 60" Fahrenheit.

Page 2

Figure 12-3 (*Continued*)

12. CONTRACT:

1) Customer must sign a separate contract for Experimental Transportation Service to each delivery point. The minimum contract term for Experimental Interruptible Transportation Service is (1) year.
2) Customer must advise the LDC 30 days in advance in writing if they wish to cancel a contract for Experimental Interruptible Transportable Service.
3) Customer is obligated to provide a copy of all contracts used to obtain and deliver natural gas to LDC's city gate, receipt points. Customer is not required to provide price information.

Page 3

Generally, applicability provisions are instituted because the LOG has determined that these requirements would be needed by a customer to benefit from a rate of this type. Also sometimes, especially on an experimental rate of this type, the LOG may want to test the rate's validity or applicability only to a certain type or class of customers.

3. *Rate*. This section sets forth the charges that are applicable in this rate. This rate is a 3-step rate design where the more natural gas used, the less expensive it becomes incrementally.
   Step #1: All usage up to 4,000 Ccf
   Step #2: All usage over 4,000 Ccf up to 20,000 Ccf
   Step #3: All usage over 20,000 Ccf

4. *Special Conditions*. This section sets forth the conditions that the customer must be responsible for to qualify for this rate.
   Condition #1: Customer must provide their own natural gas.
   Condition #2: Customer must have their own on-site natural gas replacement system.
   Condition #3: Customer agrees to curtail their use of natural gas if requested to do so by the LDC.

5. *Btu Adjustment*. This section states the method the LDC will utilize to adjust the quantities of natural gas delivered in Ccf to conform to

the thermal value of the natural gas actually used. Natural gas can be measured volumetrically (Ccf, Mcf), but it is always priced thermally (therm, Dth, MMBtu) since heat or Btu value is what the customer actually receives.

6. *Nomination*. Here is detailed how the nomination or order of the natural gas works. Everyday by 8:00 a.m., the LDC must be notified of the customer's next day's usage numbers. The usage day starts at 12:00 a.m. on the day following the 8:00 a.m. notification time. If the customer structures the contract correctly, the marketer should perform the duties required.

7. *Balancing*. This section describes what the customer's responsibility is with relation to ensuring that the quantity of natural gas nominated actually matches what is delivered. This particular rate requires balancing on a daily basis, which is generally very difficult. However, in Section #8, Item #2, there is a way out if the customer chooses to utilize this LDC's optional transportation service—for a fee, of course.

8. *Penalty Provisions*. This section outlines the various extra costs that the customer can be subject to if they do not or cannot accurately estimate their daily usages. Again, an out is provided in #2 whereby the customer can elect for an optional transportation balancing service—at a cost.
   A. Describes penalty provisions that arise ($0.50/Ccf or $5.00/Mcf) when the actual customer natural gas daily usage is either less than or greater than what was nominated (see Item #6—Nomination) by a margin of greater than 10%.
   B. Describes the provisions of another rate schedule (OTSB-142) that the customer can utilize to avoid the balancing penalty provisions of this section. The Rate Schedule OTSB-142 cited charges at a flat fee per Ccf of $0.03/Ccf ($0. 30/Mcf), which releases the customer from the imbalance penalties described in this section.

If a customer's nominated vs. actual consumption of natural gas is extremely variable (greater than ±10%), the Rate Schedule OTSB-142 might be considered to reduce and/or level out the customer costs.

Typically, a properly designed marketer contract will address the imbalance possibility and assign the risk or penalty payment to the marketer. The marketer is really only providing services and these services should include protecting the customer from delivery and usage penalties.

9 *Due Date*. Outlines the payment terms. This section sets the non-late payment charge time as 15 days after the bill processing date. This bill processing date is normally the date that the LDC actually mails the bill to the customer. It is not uncommon for an LDC bill to get delayed in various departments of a customer's bill-paying process.

Generally, an LDC bill is sent to the facility where the commodity was consumed. The facility personnel check the bill for correctness and forward it to the accounts payable department for payment. This process can take considerable time to complete; and, it is not difficult to exceed, in this case, the 15-day time limit. Many companies routinely pay late payment penalties and many times are not even aware that they are doing so.

10. *Late Payment Charge*. This entry outlines penalties on delinquent accounts. This particular LDC assesses a 2% late payment penalty per month on any unpaid amounts over a minimum $10.00 amount. This late payment fee, if compounded on an annual basis, amounts to over 24%. This fee of 2% per month is in the typical range of from 1-1/2% per month to 3-1/2% per month that most LDCs charge.

11. *Unit of Measurement*. This section describes how natural gas is measured. The unit of measurement utilized by this particular LDC is volumetric (Ccf). (All natural gas sold is based upon its thermal value.) The customer, when purchasing natural gas (and even electricity), is really buying heat (Btus). Natural gas is always sold, from point of production, in thermal units—therms, Dth, MMBtu.

Between point of production and the customer meter point, this LDC changes the unit of measurement from thermal (Btu) to volumetric (Ccf) on the monthly billing. However, if this LDC were providing the natural gas commodity to the customer, there would be a monthly true-up of the actual Btu value of the natural gas delivered during the billing month. Typically, this true-up or adjustment is called the gas cost adjustment (GCA) factor.

12. *Contract.* This section outlines the various contracts the customer must initiate before utilizing this rate. Also, detailed are the terms for cancellation by the customer. The information the customer must provide to the LDC to utilize this rate schedule includes the following items:
   A. A separate contract for each separate meter point to be served through this rate schedule (EXP-1-IT).
   B. This rate schedule is for an initial period of (1) year and continues in effect until either party cancels it. This type of contract is called Evergreen, which means that if neither party does anything after the initial contract period (1 year), the terms and conditions continue. This type of contract structure is not the best from the customer's perspective, but it is the one utilized by this LDC.

      The reason that this contract structure is not customer-friendly is that the customer must do some affirmative action to end the contract after the initial (1) year minimum period. Contracts should end automatically at a specific time as detailed in the contract and not continue for an indefinite time period until either party takes some action to end it.
   C. The customer is obligated to assure the LDC that there is a marketer contract in place that will provide the required transportation of the natural gas at the LDC's pipeline receipt points.

Even though this experimental interruptible transportation rate is relatively simple and self-explanatory, there are many items to evaluate. Knowledge of what is being done is the key to a successful LDC cost reduction strategy.

BASIC LDC DATA NEEDED— OBTAINED FROM THE LDC OR STATE REGULATORY BODY

Non-standard Rates

Non-standard rates differ from both base rates as well as experimental rates in the way they are developed and applied. Non-standard rates are negotiated generally between an LDC and a specific customer. They are generally, at least initially, discriminatory in nature and typically apply to larger customers.

Rates of this type are negotiated between the LDC and a specific customer and must be approved by the appropriate regulatory agency. Once a non-standard rate is established, it mayor may not be available for a customer that has the same usage characteristics as the customer for which the rate was originally developed.

Generally, non-standard rates are developed for large user customers that have either extremely large loads or unusual use characteristics that are not addressed adequately in a base rate structures. Most non-standard rates occur with larger LDCs and diverse customer bases.

To establish a non-standard rate is at best a drawn-out procedure. First, the LDC, as well as the regulatory agency, has to be convinced of the need for a rate of this type—no easy task in itself. Second, other LDC customers may protest a discriminatory rate since it could impact their LDC costs unfavorable—someone has to pay for the lost revenue that generally results from a non-standard rate.

Non-standard rates are not widely used or applicable to a large user base, but if a customer has a very large load or unusual usage characteristic, a non-standard rate arrangement might be applicable. Evaluate the base rate schedule in terms of usage characteristics. If large differences appear between actual usage patterns and those specified in the base schedule, the potential for a non-standard rate may exist. To determine whether non-standard rates are available, contact the appropriate regulatory agency and request a copy of any nonstandard rates that are currently available for the applicable LDC.

In natural gas, one of the most frequent uses of a non-standard rate is to persuade a natural gas customer not to leave the LDC and physically connect to an interstate pipeline. This process is called bypass and it relates to bypassing the local LDC, thereby eliminating their revenue.

In theory, *bypass* sounds good especially if a customer has an interstate pipeline under their property. In practice, not many bypass arrangements are ever completed because generally the LDC will negotiate a favorable rate with a customer.

The general feeling by most LDCs is that some revenue is better than no revenue. In these types of cases, the LDC will work with the customer and the regulatory commission to structure a rate that will keep the customer with the LDC and allow at least, some revenue from the customer.

What a typical non-standard rate of this type might look like is shown in Figure 12-4.

Figure 12-4. Example of a Non-standard Rate

Non-standard Rate CSR-1 (Competitive Service Rider)

1. APPLICABILITY:
Applicable at company's discretion to commercial and industrial customers that have natural gas requirements, which are subject to competition from sources other that the company. Competition is considered to exist when a customer is located in the company's service territory and has the ability to obtain its natural gas requirements from a supplier not regulated by the state regulatory agency.

2. RATE:
Standard non-standard rates apply except that the level of the natural gas and/or transportation charges may be reduced for each customer based upon a consideration of customer's usage characteristics and the lowest cost natural gas and/or transportation available to the customer from a source other than company.

3. TERMS AND CONDITIONS OF SERVICE:
   A. Customer must provide company with information that documents the customer is not likely to purchase natural gas and/or transportation from the company under any non-standard rate available from the company.
   B. Minimum usage that can be served through this rider is 3,000 Dth per month.
   C. Customer must execute a natural gas and/or transportation agreement with the company, which will include:

      1) The minimum rate under this rider, which will recover at least the incremental cost of providing service including the cost of natural gas and/or transportation
      2) The maximum possible rate reduction under this rider will not exceed the difference between the applicable non-standard rate and the cost to the customer of the lowest cost competitive natural gas and/or transportation supply
      3) The term of service under this rider—not less than (1) year and not longer than (5) years

Page 1

Figure 12-4 (*Continued*)

4) The size of the load served under this rider
5) An annual minimum charge to fully recover distribution and/or transportation costs

4. The company, within a general rate case, is allowed to seek recovery of the difference between the usual non-schedule rate and the rider rate, times the usage level during the test year period.
5. A rate under this rider shall meet the conditions of the regulatory agency statutes for other customers in this same customer class.
6. Unless the regulatory agency determines that it would be in the public interest, a rate under this rider shall not compete with other intrastate natural gas companies whether For-profit, Municipal, or Cooperative.
7. A rate offered under this rider may not be offered to a customer in whom the company has a financial interest greater than 50%.

Page 2

Analysis of Non-standard Rate—CSR-1

This rate allows an LDC to compete with a third party for the customer's natural gas and/or transportation business. This third party, in most instances would be the interstate pipeline currently serving the intrastate LDC. A transaction of this type would be called a *bypass*.

1. *Applicability*

This section addresses the type of customers and conditions which must be present before these customers can be considered for inclusion under this rider.

Since a rate of this type is inherently discriminatory, the LDC can, within this rider's written boundaries, choose which, if any, individual customers it will offer this ratemaking process. The discrimination factor is minimized since any LDC shortfalls caused by revenue discounts offered through this type of ratemaking process normally cannot be passed through to any other customer. Also, the offering of this type of discount can preserve at least some positive revenue, which can help other customer classes by potentially reducing LDC costs that ultimately affect overall costs.

2. *Rate*

This section details the maximum rate reduction that a customer can obtain. Generally, the rate reduction cannot result in a cost to the customer of less than what would result if the customer opted for the competitor's rate. There are no specific rate figures given here since each customer's situation will be different and as such, requires differing levels of cost reductions or concessions by the LDC.

3. *Terms and Conditions of Service*
   This section lists the various conditions that apply to this rate:
   A. In order for a customer to obtain this rate, they must provide evidence substantiating the possibility of their being able to obtain a less costly competitive rate. Generally, this evidence must include any up-front costs the customer will incur in obtaining the competitive rate. In the case of bypass costs, these might include:
      1) Installation of a pipe connecting the customer facility to the interstate pipeline
      2) Installation of the tap on the interstate pipeline as well as pressure reduction equipment
      3) Any miscellaneous items that might be present

   B. This section states the minimum monthly usage the customer must have before they can even apply for this rate. Generally, the LDC will evaluate a historical 1- to 5-year period in determining whether the customer qualifies. If customer multiple meters are present, it is important that all meter usages are totaled since an actual bypass would provide service to all customer meter points.

   C. This section outlines the written agreement conditions which the customer must sign before this rate can be utilized:
      1) This defines the minimum revenue the LDC will realize from a customer on this rate. The LDC will recover at least their incremental or actual cost of the natural gas and/or transportation to the customer's meter point.
      2) This states the maximum discount available to the customer. The LDC will not price its natural gas and/or transportation lower than what the customer would pay for the competitive service, even if this cost is higher than the LDC's incremental or actual cost.
      3) This outlines the minimum and maximum contract terms for

this rate.

4) This defines the size of the natural gas and/or transportation load to be included in this rate.

5) This states that in addition to the provisions in A, and B., the LDC will at least recover its cost of distribution of natural gas and/or its cost of transportation.

4. This section states that in the course of a general rate case (*not a special filing for this rider*) the LDC will have the right to attempt to recover any lost revenue from the utilization of this rider from the general rate base.

   This can cause a problem because of what it is indicating to all customers of the LDC. If the LDC cannot realize its normal margin from a customer utilizing this rider, all other customers must subsidize this shortfall. Not many customers embrace this philosophy enthusiastically.

   However, the truth of the matter is that if the customer or customers utilizing this rider were to actually leave the LDC system, the negative impact on all other customers would be even more severe because of ongoing LDC costs that would not be recovered by the LDC from the customer or customers that left the system.

   Even though the customer or customers that utilize this rate do have a lower rate than perhaps does any other customer of the LDC, the LDC is still recovering their cost of natural gas as well as distribution and/or transportation costs [Item #3, 1) and 5)].

5. This section states that general conditions that apply to all customers will also apply to customers utilizing this rate. The only LDC concession is in the cost of the natural gas and/or transportation.

6. This section says that the LDC will not compete with other intrastate LDCs unless the regulatory agency deems it to be in the general public interest to do so. This provision is a standard declaration and is included to keep various LDCs within a given intrastate area from competing with each other which in theory, could lead to the ultimate harm of all LDCs in that area.

   Is *no competition* in the general public good? The answer to this depends upon who is involved. If I am an LDC, the answer is probably yes. If I am a customer, the answer is probably no. *Protectionism* as opposed to *competition* has never been prone to induce efficiency or least-cost scenarios.

7. This section limits the financial interest an LDC can have in a customer's business that is applying for this rate. This is a general clause that is used since if the LDC owned a predominate portion of a company applying for this rate, the concession could be conceived as benefiting the LDC indirectly at the expense of other ratepayers.

BASIC LDC DATA NEEDED—
OBTAINED FROM THE LDC OR STATE REGULATORY BODY

Analysis of Rebate Programs

LDCs may have programs to encourage the use of natural gas. Also, LDCs may offer rebate programs that encourage customers to utilize their natural gas efficiently by paying for or providing rebates for those items that are more efficient in natural gas usage. Rebate programs range from *not worth much* to *extremely beneficial*.

These programs change frequently and sometimes a specific amount of money is allocated for a program which means—when the money is gone, the program is ended. If an LDC has programs to encourage the use of natural gas or a rebate program, they will also have an in-house specialist that can be utilized for an on-site evaluation of a facility to determine the applicability of the LDC program to a particular situation. The LDC service representative can provide program information as well as arrange for an on-site evaluation by the LDC specialist.

These programs typically include the following items and/or processes, although not all items are included in all programs.

Programs to Encourage the Use of Natural Gas

A. Audits by the LDC to determine applicability of natural gas usage
B. Assistance in converting air conditioning to natural gas
C. Assistance in retrofitting manufacturing processes to natural gas
D. Assistance in converting from purchased steam to self-generated on-site steam by natural gas
E. Assistance in installing natural gas-fired distributed generation systems
F. Assistance in installing natural gas engine-operated chillers

G. Custom natural gas programs structured to individual customer requirements

These programs are individually negotiated on a customer and LDC basis.

Rebate Programs

A. Audits by the LDC to determine rebate applicability
B. Installation of efficiently designed burners
C. Assistance in installing on-site, alternate fuel, backup systems so that a customer can reduce their natural gas costs
D. High efficiency natural gas-fired heating systems
E. Industrial and commercial natural gas cooling systems
F. Weatherization programs
G. Heating systems for process drying
H. Industrial refrigeration
I. High efficiency heat exchangers
J. Programs structured to individual customer requirements

These programs are individually negotiated on a customer and LDC basis.

UNREGULATED MARKETING AFFILIATE (UMA) PROGRAMS

Some LDCs currently have capacity distribution constraints. This means that they may have periods of distribution shortfall during some period of a 24-hour day. The LDC can do several things to compensate for this shortfall. They can construct new pipeline capacity that is very expensive, or they can offer their customers financial incentives to reduce natural gas usage during the LDC's shortfall periods.

Many LDCs, through unregulated marketing affiliates (UMA), offer programs that encourage customers to reduce their natural gas needs by paying for, or providing incentives for, items that reduce the LDC's distribution capacity problems. These programs change frequently and sometimes a specific amount of money is allocated for a program, which means that when the money is gone, the program is ended. However, these programs are generally worth at least investigating.

Generally, if a marketing affiliate has a program, they will also have an in-house specialist that can be utilized for an on-site evaluation of a facility to determine the applicability of the program to a particular situation. The LDC service representative can provide program information as well as arrange for an on-site evaluation by the appropriated specialist.

Why is Reducing Natural Gas Costs so Complicated?

As with anything new, unfamiliarity makes the process, or procedure, seem more difficult than it really is. The base data collection portion of reducing natural gas costs is probably the most important part of the undertaking. Once the process of getting the information outlined in this chapter begins, it becomes apparent that only two entities need to be contacted—the LDC and the state regulatory agency.

SYNOPSIS—REDUCING NATURAL GAS COSTS

1. **Customer's Responsibility**
   It is the customer's responsibility to develop a cost reduction strategy for natural gas

2. **Evaluate the Basic Rate Data Needed**
   a) Applicable rate schedule
   b) Alternative rate schedules
   c) Experimental rate schedules
   d) Non-standard rate schedules
   e) Rebate programs

3. **Evaluate State Data Needed**
   a) Sales tax statutes
   b) Economic development and enterprise zone programs

4. **Evaluate Unregulated Marketing Affiliate (UMA) Programs**

Chapter 13

The Natural Gas Flow Process

THE NATURAL GAS LDC BILLING PROCESS— FROM EXPLORATION TO CUSTOMER BURNER TIP

Finding Natural Gas

An important point that needs to be understood is how natural gas flows—the process of getting natural gas from where it originates to where it is used. This chapter explains the natural gas process from wellhead to customer meter point.

EXPLORATION

Finding Where the Natural Gas is Located

The practice of locating natural gas deposits has been transformed dramatically with the advent of advanced technology.

Early in the industry, the only way of locating underground natural gas deposits was to search for surface evidence of these underground formations. The only evidences of any underground deposits were the seepages of oil or emissions of natural gas.

However, because such a low proportion of natural gas deposits actually seep to the surface, the exploration process became very inefficient and difficult. As demand for fossil fuel energy has increased dramatically over the past years, so has the necessity for more accurate methods of locating these deposits.

Sources of Data

Technology has allowed an increase in the success rate of locating natural gas reservoirs. Geologists and geophysicists use state-of-the-art technology to gather data that can later be interpreted and used to make educated guesses as to where natural gas deposits exist.

However, the process of exploring for natural gas deposits incorporates uncertainty due to the complexity of searching for something that is often thousands of feet below ground.

Geological Surveys

By surveying and mapping surface and subsurface characteristics in a certain area, the geologist can determine which areas are most likely to contain natural gas reservoirs. To assist in this determination, the geologist investigates outcroppings of rocks on the surface in valleys and gorges, plus rock cuttings and samples obtained from digging of irrigation ditches, water, wells, and other natural gas wells.

This information is all combined to allow the geologist to ascertain the fluid content, porosity, permeability, age, and formation sequence of the rocks underneath the surface of a particular area. For example, a geologist may study the outcroppings of rock to gain insight into the geology of the subsurface areas.

Once it has been determined that it is geologically possible for a natural gas or petroleum formation to exist, further tests are performed to verify preliminary data. These tests allow a more accurate simulation of underground formations that are commonly associated with natural gas and petroleum reservoirs. These tests are commonly performed by a geophysicist—one who uses technology to find and map underground rock formations.

EXTRACTION

Getting the Natural Gas out of the Ground

Once a potential natural gas deposit has been located, an actual drilling to where the natural gas is thought to exist will take place. Although the process of digging deep into the earth's crust to find deposits of natural gas that mayor may not actually exist is expensive, the industry has various innovations and techniques, which decrease cost and increase efficiency of drilling for natural gas. The advancement of technology has contributed greatly to increased efficiency and success rate for drilling natural gas wells.

If natural gas is found in a new well, it is developed and designated a *productive well* and will be completed. If the exploration process does not find marketable quantities of natural gas, the well is termed a *dry well* and further efforts are discontinued.

PROCESSING NATURAL GAS

Making the Natural Gas Usable

Customer meter point natural gas is much different from the natural gas that is present in wells prior to extraction. The processing of natural gas, in many respects, is less complicated than the processing and refining of crude oil, but processing is required before the natural gas can be used by the ultimate customer

The natural gas used by customers is composed almost entirely of methane. Even though natural gas found at the wellhead still is composed primarily of methane, it is by no means pure. Raw natural gas comes from three types of wells:

1) Oil wells
2) Gas wells
3) Condensate wells

Natural gas that comes from oil wells is typically termed *associated gas*. This gas can exist separate from oil in the formation *free gas*, or dissolved in the crude oil *dissolved gas*.

Natural gas from gas and condensate wells, in which there is little or no crude oil, is termed *nonassociated gas*. A gas well typically produces raw natural gas by itself, while condensate wells produce free natural gas along with a semi-liquid hydrocarbon condensate. Whatever the source of the natural gas, it commonly exists in mixtures with other hydrocarbons; principally ethane, propane, butane, and pentanes. Also, raw natural gas contains water vapor, hydrogen sulfide, carbon dioxide, helium, nitrogen, and other compounds.

Natural gas processing consists of separating all of the various hydrocarbons and fluids from the pure natural gas, to produce what is known as *pipeline quality* dry natural gas. Interstate pipelines impose restrictions on the composition of the natural gas that is allowed into the pipeline. Natural gas must be purified before it can be transported.

While some of the needed processing can be accomplished at or near the wellhead *field processing*, the complete processing of natural gas takes place at a processing plant, usually located in a natural gas producing region.

The extracted natural gas is transported to these processing plants through a network of gathering pipelines, which are small-diameter, low-pressure pipes. A complex system can consist of thousands of miles of pipes, interconnecting the processing plant to 100 or more wells in a gathering area.

The actual practice of processing natural gas to pipeline dry gas quality levels can be quite complex, but usually involves at least the processes shown following to remove the various impurities:

1. Oil and condensate removal
2. Water removal
3. Separation of natural gas liquids

Oil and Condensate Removal

The actual process used to separate oil from natural gas, as well as the equipment that is used, can vary widely. Although dry pipeline quality natural gas is virtually identical across different geographic areas, raw natural gas from different regions may have many different compositions and separation requirements.

In many instances, natural gas is dissolved in oil underground primarily due to the pressure that the formation is under. When this natural gas and oil is produced, it is possible that it will separate on its own simply due to decreased pressure. In these cases, separation of oil and gas is relatively easy as the two hydrocarbons are sent in separate paths for further processing.

Water Removal

In addition to separating oil and some condensate from the wet gas stream, it is necessary to remove most of the associated water. Most of the liquid is removed by simple separation methods at or near the wellhead. However, the removal of the water vapor that exists in natural gas requires a more complex treatment. This treatment consists of dehydrating the natural gas, which usually involves one of two processes—either *absorption* or *adsorption*.

1. **Absorption** occurs when the water vapor is taken out by a dehydrating agent

2. **Adsorption** occurs when the water vapor is condensed and collected on the surface of a collector assembly

Separation of Natural Gas Liquids

Natural gas coming directly from a well contains many natural gas liquids that are commonly removed. In most instances, natural gas liquids (NGLs) have a higher value as separate products and are economical to remove from the gas stream.

The removal of natural gas liquids usually takes place in a relatively centralized processing plant and uses techniques similar to those used to dehydrate natural gas. There are two basic steps to the treatment of natural gas liquids in the natural gas stream:

1. Liquids are extracted from the natural gas
2. Natural gas liquids are separated down to the base components

INTERSTATE TRANSPORTATION

Getting the Natural Gas to Where it is Needed

The efficient movement of natural gas from producing regions to consumption regions requires an extensive and elaborate transportation system. In many instances, natural gas produced from a particular well will have to travel a great distance to reach the ultimate customer.

The transportation system for natural gas consists of a complex network of pipelines, which are designed to quickly and efficiently transport natural gas from its origin to areas of high natural gas demand. Transportation of natural gas is closely linked to its storage facilities. Should the natural gas being transported not be needed at that time, it can be put into storage facilities for when it is needed. There are essentially three major types of distribution along the transportation route:

1. Gathering system (interstate/intrastate)
2. Interstate pipeline
3. Intrastate system

The Gathering System

The gathering system consists of low pressure, small diameter pipelines that transport raw natural gas from the wellhead to the processing plant. These pipelines are typically intrastate or within a given state.

Pipeline Classification

1. **Interstate Pipelines**

   Interstate pipelines carry natural gas across or between state boundaries and are subject to the Federal Energy Regulatory Commission (FERC).

2. **Intrastate Pipelines**
   Intrastate pipelines transport natural gas within a particular state and are subject to the state regulation agencies where the pipelines are located.

Interstate Natural Gas Pipelines

The interstate pipeline network transports natural gas from processing plants to those areas with natural gas requirements, particularly large, populated urban areas.

Interstate pipelines are the highways of natural gas transmission. Natural gas is transported through interstate pipelines at pressures from 200-1500 pounds per square inch (psi). This reduces the volume of natural gas being transported and provides the pressure to move the natural gas through the pipeline.

Pipeline Components

Interstate pipeline systems consist of a number of components each of which ensure the efficiency and reliability of the system.

Pipes

Pipelines are from 1-52 inches in diameter. The smaller diameter pipes are used in gathering and distribution systems. Mainline pipes, the principle pipeline in a given system, are usually between 16-52 inches in diameter.

Compressor Stations

Natural gas is pressurized as it travels through an interstate pipeline. To ensure that the natural gas flowing through anyone pipeline remains pressurized upstream, compression is required periodically along the pipe.

This is accomplished by compressor stations, usually placed at 40-100 mile intervals along the pipeline. The natural gas enters the compressor station, where it is compressed by either a turbine or reciprocating engine. The compression stations consume pipeline natural gas to operate. This usage causes what is called *system shrink*—less natural gas at the end of the pipe than at the entrance of the pipe.

In addition to compressing natural gas, compressor stations also usually contain some type of liquid separator, much like the ones used during its processing. Usually, these separators consist of scrubbers and filters that capture any liquids or other undesirable particles from the natural gas in the pipeline.

Although natural gas in pipelines is considered dry gas, it is not uncommon for a certain amount of water and hydrocarbons to condense out of the gas while in transit. The liquid separators at compressor stations ensure that the natural gas in the pipeline is as pure as possible.

Metering Stations

In addition to compressing natural gas to reduce its volume and push it through the pipe, metering stations are placed periodically along interstate natural gas pipelines. These stations allow pipeline companies to monitor and manage the natural gas in their pipes. Essentially, these metering stations measure the flow of gas along the length of the pipeline.

Valves

Interstate pipelines include a great number of valves along their entire length. These valves work like gateways—usually open to allow natural gas to flow freely, or closed to stop gas flow along a certain section of pipe.

There are many reasons why a pipeline may need to restrict gas flow in certain areas. For example, if a section of pipe requires replacement or maintenance, valves on either end of that section of pipe can be closed to allow engineers and work crews safe access. These large valves are generally placed every 5-20 miles along the pipeline.

Intrastate Pipelines

The intrastate natural gas pipeline network begins at the city gate LDC (local distribution company) meter point and ends at the customer meter point. In many respects, the intrastate pipeline is similar to the interstate network—there can be compressor and metering stations together with numerous valving locations.

One of the large differences between interstate and intrastate is in the regulatory oversight responsibility. *Interstate* pipelines are regulated by FERC. *Intrastate* pipelines are regulated by state regulatory agencies. The big difference between interstate and intrastate regulation is:

1. Interstate pipelines have only one regulatory agency—FERC
2. Intrastate pipelines have 50 different state regulatory agencies

Physically interstate and intrastate pipelines are similar in their structures and operations.

NATURAL GAS STORAGE

Having Natural Gas When and Where it is Needed

Natural gas, like some other commodities, can be stored for an indefinite period of time. The process of exploration, production, and transportation of natural gas takes time. The natural gas that reaches its destination is not always needed right away, so it is injected into underground storage facilities. These storage facilities can be located near market centers that do not have a ready supply of locally produced natural gas.

Natural gas storage plays a vital role in maintaining the reliability of supply needed to meet the demands of customers. Historically, when natural gas was a completely regulated commodity, storage was part of the bundled product sold by the pipelines to LDCs.

Storage is used to serve only as a buffer between transportation and distribution to ensure adequate supplies of natural gas are in place for seasonal demand shifts and unexpected demand surges. In addition to serving those purposes, natural gas storage is also used by industry participants for commercial reasons—storing natural gas when prices are low, and withdrawing and selling natural gas when prices are high.

Base Load vs. Peak Load Storage

There are basically two uses for natural gas in storage facilities:

1. Meeting base load requirements
2. Meeting peak load requirements

There are three reasons for natural gas storage:

1. Meeting season demand requirements
2. As insurance against unforeseen supply disruptions
3. Hedging against future price escalation

Base load storage capacity is used to meet seasonal demand increases. Base load facilities are capable of holding enough natural gas to satisfy long-term seasonal demand requirements.

Typically, the turnover rate for natural gas in these facilities is on a yearly basis. Natural gas is generally injected during the summer (non-heating season), which usually runs from April through October and withdrawn during the winter (heating season), usually from November to March.

While the reservoirs are larger, the delivery rates are relatively low, which means the natural gas that can be extracted each day is limited. Beneficially, these facilities provide a prolonged and steady supply of natural gas. Depleted gas reservoirs are the most common type of base load storage facility.

Peak load storage facilities, on the other hand, are designed to have high-deliverability for short periods of time, which means natural gas can be withdrawn from storage quickly should the need arise. Peak load facilities are intended to meet sudden, short-term demand increases. These facilities cannot hold as much natural gas as base load facilities, but can deliver smaller amounts of natural gas more quickly and be replenished in a shorter amount of time than base load facilities.

While base load facilities have long-term injection and withdrawal seasons—turning over the natural gas in the facility about once per year, peak load facilities can have turn over rates as short as a few days or weeks. Salt caverns are the most common type of peak load storage facility, although aquifers may be used to meet these demands as well. Natural gas is usually stored underground, in large storage reservoirs. There are three main types of underground storage:

1. Depleted natural gas reservoirs
2. Aquifers
3. Salt caverns

In addition to underground storage, natural gas can be stored as liquefied natural gas (LNG). LNG allows natural gas to be shipped and stored in liquid form, meaning it takes up much less space than natural gas in its normal state.

Depleted Gas Reservoirs

The most prominent and common form of underground storage consists of depleted gas reservoirs. Depleted reservoirs are those formations that have already been tapped of all their recoverable natural gas. This leaves an underground formation that is geologically capable of holding natural gas. Also, using an already developed reservoir for storage purposes allows the use of the extraction and distribution equipment left over from when the field was productive.

Having this type of reservoir extraction network in place reduces the cost of converting a depleted reservoir into a storage facility. Depleted reservoirs are also attractive because their geological characteristics are already well known. Of the three types of underground storage, depleted reservoirs, on average, are the cheapest and easiest to develop, operate, and maintain.

Aquifers

Aquifers are underground porous, permeable rock formations that act as natural water reservoirs. In certain situations, these water containing formations may be reconditioned and used as natural gas storage facilities. As they are more expensive to develop than depleted reservoirs, these types of storage facilities are usually used only in areas where there are no nearby depleted reservoirs. Traditionally, these facilities are operated with a single winter withdrawal period, although they may be used to meet peak load requirements as well.

Salt Caverns

Underground salt formations offer another option for natural gas storage. These formations are well suited to natural gas storage in that salt caverns, once formed, allow little injected natural gas to escape from the formation unless specifically extracted. The walls of a salt cavern also have the structural strength of steel, which makes it very resilient against reservoir degradation over the life of the storage facility. The typical costs associated with natural gas storage facilities are:

1. *Injection*—per unit cost
   (Injecting the natural gas into storage)

2. *Storage*—per unit cost
   (Holding the natural gas in the storage facility)

3. *Shrinkage Factors*—percentage cost
   (Loss of natural gas in storage facility due to migration of natural gas molecules through the storage facility walls, floor, and ceiling)

4. *Extraction*—per unit cost
   (Removing the natural gas from the storage facility)

Although there are at least four separate costs associated with storing natural gas, they generally are minimal when compared to the benefits that can be gained by the process.

LDC DISTRIBUTION PROCESS

Getting the Natural Gas to the Customer

Distribution is the final step in delivering natural gas to the customer. While some large industrial and commercial customers receive natural gas directly from high capacity interstate and intrastate pipelines (usually contracted through natural gas marketers), many customers receive natural gas from a local distribution company (LDC).

LDCs are companies involved in the delivery of natural gas to customers within a specific geographic area. There are two basic types of local distribution companies:

1. Those owned by investors
2. Those owned by local governments

LDCs typically transport natural gas from delivery points along interstate pipelines through thousands of miles of small-diameter distribution pipe. Delivery points to LDCs, especially for large municipal areas, are often termed *city gate* and are important market centers for the pricing of natural gas.

Usually, LDCs take ownership of the natural gas at the city gate and deliver it to each individual customer's location of use. This requires an extensive network of small-diameter distribution pipe. It has been estimated that over one million miles of distribution pipe exist in the United States.

Delivery of Natural Gas

The delivery of intrastate natural gas to the customer meter point by a LDC is much like interstate transportation of natural gas. However, intrastate distribution involves moving smaller volumes of gas at much lower pressures over shorter distances to a greater number of individual users. Small-diameter pipe (1-6 inches) is generally used to transport natural gas from the city gate to individual customers.

The natural gas is periodically compressed to ensure pipeline flow, although local compressor stations are typically much smaller than those used for interstate transportation. Because of the smaller volumes of natural gas to be moved, as well as the small-diameter pipe that is used, the pressure required to move natural gas through the distribution network is much lower than that found in the transmission pipelines.

While natural gas traveling through interstate pipelines may be compressed to as much as 1,500 pounds per square inch (psi), natural gas traveling through the distribution network maybe as low as 3 psi of pressurization.

The natural gas to be distributed also is scrubbed and filtered to ensure low moisture and particulate content, even though it has already been processed prior to distribution through interstate pipelines.

In addition, *Mercaptan*—the source of the familiar rotten egg smell in natural gas—is added by the LDC prior to distribution. This is added because natural gas is odorless and colorless, and the familiar odor of Mercaptan makes the detection of leaks much easier.

Regulation of Natural Gas Distribution

Traditionally, LDCs have the exclusive rights to distribute natural gas in a specified geographic area, as well as perform services like billing, safety inspection, and providing natural gas hookups for new customers. Like interstate pipelines, LDCs are monopolistic in nature. Because of the cost of implementing the distribution infrastructure, it would be uneconomic to lay overlapping distribution networks in anyone area, meaning that in most areas there is only one LDC offering distribution services.

CLASSES OF NATURAL GAS SERVICE

Types and Classes of Natural Gas Service

Previously discussed, there is only one quality or type of natural gas—pipeline quality. However, there are several different classes of natural gas service. These classes of service are, or should generally be, available form the intrastate natural gas LDC. These classes of service are:

1. Firm Service
2. Interruptible Service
3. Flex or Adjustable Rate Service
4. Firm Transportation Service of Customer Natural Gas
5. Interruptible Transportation Service of Customer Natural Gas
6. LDC Marketer Service (commodity purchase by customer)

These six types of service all provide the same quality of natural gas, but costs can vary greatly because of the differences in delivery categories.

Firm Service Natural Gas (Glass 1) will be the most expensive and Interruptible Transportation (Glass 5) will generally result in the least expensive natural gas for the customer.

All of the natural gas delivered in any on the six categories detailed will be the same and normally will originate in the same natural gas gathering zones, travel through the same interstate and intrastate pipelines, and end up at the same burner tip meter point on the customer's property. Any and all of these services originate with the intrastate natural gas LDC.

Some LDCs offer all six classes of service and some only offer one or two. If a customer is served by a LDC that offers only Firm Service, or Firm and Interruptible Service, they are probably not able to realize the lowest cost natural gas service that could be available.

On practically all interstate natural gas pipelines, transportation of customer natural gas is available. However, if a customer's intrastate natural gas LDC does not offer a Firm or Interruptible Transportation Service, then the customer cannot take advantage of the interstate transportation service. In situations like this, the intrastate natural gas LDC customer can intervene in natural gas rate cases and request/require/ encourage the LDC to provide customer-friendly transportation types of service.

EXPLANATION OF THE SIX CLASSES OF NATURAL GAS SERVICE

Firm Service

Natural gas that is purchased under *Firm* class of service is typically for a type of customer that cannot have their natural gas supply interrupted or curtailed. Generally, the type of natural gas customer that would choose this class of service would not have an alternate or backup fuel source that could be utilized in case of interruption of natural gas.

This being the case, the customer will pay a premium for not being subject to interruptions of their natural gas supply. Many times a customer of this type could afford to install an on-site alternate or backup fuel source, which would be paid out of the differential between Firm and Interruptible or LDC transportation classes of service cost.

Interruptible Service

Interruptible class of service is utilized by those customers who can either accommodate interruptions of their natural gas supply or who have on-site alternate or backup fuel sources. If a customer currently could be in this classification, they could reduce their natural gas costs by changing to interruptible transportation if it is available from the LDC.

Many times, even the mention of the word *interruption* instills fear in the heart of a customer because they envision multiple long-term periods of interruptions of natural gas. When evaluating Interruptible Service, utilize data to assist in the decision-making process. Generally, the data to use is available from the LDC and is as follows:

1) History of number of natural gas interruptions for the last five years
2) History of natural gas interruption lengths (days, hours) for the last five years
3) History of when natural gas interruptions occurred in the last five years (year, month, week, day, hour)
4) LDC's evaluation and estimation of current or future interruptions considering system load, customer distribution (residential, commercial, industrial)

Once a customer has available the data listed in these four items, they can make an informed decision as to the applicability of Interruptible Service to their particular situation.

Flex or Adjustable Rate Service

Flex or *Adjustable Rate Service* is available to a customer who has the ability to switch from LDC-supplied natural gas to their own fuel source (propane, fuel oil, etc.). If a customer who has this switching ability elects to discontinue natural gas service generally due to natural gas being more expensive than their own fuel source, the LDC ends up with no revenue since no natural gas flows to the customer.

When a customer is inclined to discontinue natural gas usage due to cost considerations, the LDC will offer to sell natural gas to the customer on a flexible pricing basis (Flex Rate) that is no more costly than what the customer's own fuel would be. The customer usually would rather utilize LDC natural gas anyway, so both the LDC and the customer are satisfied. Although the LDC realizes less revenue in this type of arrangement, they still retain the natural gas user as a customer.

A Flex Rate type of service generally works this way: on a monthly basis, the customer provides an affidavit to the LDC that details the actual cost of their own fuel source. The LDC then evaluates this customer's fuel cost and, if practical, will match the cost to retain the customer as a natural gas user.

Firm Transportation Service

Firm Transportation Service is utilized when a customer wants to arrange for their own natural gas supplies and utilize the LDC only for transportation of natural gas.

By utilizing the Firm class of transportation, the customer is not required to have available on-site alternate or backup fuel supplies. In the event that the customer's natural gas is physically interrupted, the LDC will supplement the customer's natural gas with LDC system supply natural gas. There are more and more intrastate natural gas LDCs offering this class of service to their transportation customers. However, one problem that seems to occur frequently is that the cost differential between Firm and Interruptible classes of service is so large that opting for the Firm service is not financially viable.

When determining whether to utilize Firm or Interruptible classes of service, the cost differential between the two classes must be evaluated. In some instances, it may be cost-effective to install on-site alternate or back-up fuel sources and use the Interruptible Transportation Service.

Interruptible Transportation Service

Interruptible Transportation Service is the least expensive that an intrastate natural gas LDC will have available. As with the Firm Transportation Service, the only thing the customer is purchasing from the LDC is transportation of customer natural gas through the intrastate natural gas LDC's pipeline.

When Interruptible Transportation class of service is utilized, the customer will have to provide their own alternate or backup fuel source since the LDC is not providing natural gas but transportation services only.

Should a customer have their natural gas supply interrupted and continue to utilize LDC system supply natural gas, a large penalty will usually be assessed by the LDC. This penalty may be in the range of $50 or more per unit (Dth, Mcf, MMBtu) of LDC system supply natural gas used.

In addition to this penalty, the LDC will charge their normal Firm Natural Gas Service Rate for all of the LDC system supply natural gas used.

Select this class of service only when an actual interruption of natural gas can be accommodated or where on-site alternate or backup fuel supplies exist.

LDC Marketer Service

When a natural gas LDC offers Marketer Service, they are acting as the customer's agent to obtain non-LDC supplied natural gas. They are performing the same function as an independent marketer would provide for the customer.

The main advantage an *LDC Marketer Service* seems to have over an independent marketer is the comfort level the customer might have since the LDC may seem to be more utility related and as such, perhaps more trustworthy or less likely to cause customer problems.

However, a LDC marketer is the same as and performs the same way as an independent marketer. Before a LDC can institute a Marketer Program, the state regulatory agency will require the Marketer section of the LDC to be a completely independent entity with no direct relationship with the LDC. This is done so that a conflict of interest situation does not arise where the LDC might tend to favor their marketer over an independent one.

If the correct process to select a marketer is utilized, then a LDC vs. non-LDC marketer can be objectively evaluated, then the one which is best suited to the customer's needs can be utilized.

ALTERNATE/BACKUP FUEL SYSTEMS

The least expensive natural gas that will be available to a customer will be where the customer utilizes a marketer to purchase and a LDC to deliver the customer's natural gas. The least expensive procedure of this type is where both the natural gas and the delivery are on an interruptible basis.

To be able to utilize this type of interruptible service, a customer must either be able to curtail natural gas usage when required, or have on-site alternate or backup fuel sources available. To fully understand what is involved in these on-site fuel sources, the following information is presented.

Alternate Fuels

Alternate fuels (propane, fuel oil, coal, wood) are those fuels that can be used as a replacement or supplement to natural gas as a source of energy. The most widely used sources are either liquefied petroleum (propane) or fuel oil.

There currently are LDCs that require an in-place operable alternate fuel source before they will allow Interruptible Service. Not only might an alternate fuel source result in lower natural gas costs, but it also provides the availability of a backup in the event of curtailment. Note: a curtailment can happen even when using Firm Service natural gas. Depending upon the use of the natural gas, generally an on-site propane air-backup unit is the most flexible and cost effective to install and use.

Many LDCs offer reduced natural gas rates to large customers in exchange for an agreement to go offline during periods of peak demand. Generally, a firm commitment to supply natural gas is replaced by best efforts to supply under one or more interruptible options.

Customers can often save from 10% to 30% or more by choosing Interruptible Service. For this service to be viable, most customers need an alternate fuel. The ideal solution would be a fuel that can be burned with no change in equipment and operations. That is where propane-air backup systems bring unique advantages to the equation.

Commercial propane is one of several liquefied petroleum gasses (LPCs) derived from processing natural gas and crude oil. Stored as a liquid under pressurized propane (G_3H_8) vaporizes easily to produce a gas.

As a gas, propane is heavier and has a higher energy density than methane (GH_4), the primary component of natural gas. Consequently, some adjustment is almost always required to burn propane in equipment setup for natural gas.

Spark-ignited natural gas (SING) technology, similar to LDC peak-shaving plants, a customer-owned propane-air backup system avoids this requirement by blending propane with air to produce a mixed gas with burning characteristics similar to natural gas. This means gas energy keeps flowing when the natural gas supply is shut off with no adjustments to combustion equipment.

Various grades of fuel oil are also commonly used as alternate fuels, especially in boilers. The basic problem many customers face is that their gas fuel use may not be limited to boilers. Boilers can easily be equipped with oil/gas burners, but gas-fired rooftop ventilation equipment, process ovens, and a host of other equipment cannot. This is where propane-air systems shine. Propane-air provides an alternate fuel that is transparent to fuel-burning equipment. In fact, the switchover can even occur while the equipment is being operated.

Even for boiler plants, propane-air systems have advantages over oil systems:

1) Oil storage is subject to stringent requirements to avoid environmental damage due to leakage
2) Changeover from gas to oil in large boilers often involves changing the burner nozzle, a step that can require hours for the boiler to cool before the change can be performed
3) Maintaining oil fuels is difficult as oil quality degrades over time

Propane-air as an alternate fuel avoids these problems and preserves the low level of air emissions associated with natural gas combustion.

SYNOPSIS—THE NATURAL GAS FLOW PROCESS

Knowing how natural gas moves from where it is produced to the location of the customer's meter point, is basic to the natural gas cost reduction process. Understanding the flow process allows a customer to accurately calculate natural gas transportation costs and the part they have in the overall cost structure.

Chapter 14

The Natural Gas Billing Process

THE BASICS

To be able to accurately analyze a LDC billing, an understanding of the items or billing components is a must. Since there is almost no uniformity in the format of LDC billings, the basic items described in this chapter may appear in any sequence on a given billing but will generally be present somewhere on the actual document.

It is very important that the following items described are thoroughly understood since they will provide the basis for investigation of natural gas costs as shown on the natural gas billing. When analyzing an actual natural gas billing for the presence of these items, problems may be encountered in locating them. If this happens, contact the LDC service representative for information concerning where or how the item in question is addressed.

How to Analyze Your Natural Gas Billing

Since practically all LDCs utilize different styles of billing formats, as well as types of information displayed on billings, there is no single fits all example. In view of the fact that there is not a single example that would apply to the billing design everyone, there are three examples shown:

1. Figure 14-1—LDC Billing Example
2. Figure 14-2—LDC Delivery Service Billing
3. Figure 14-3—Marketer Natural Gas and Delivery Service Billing

While these examples might be somewhat different from any specific billing, they do contain all major categories of data that would normally be found on any actual billing. Once these examples and the individual items are examined, any actual billing will be much easier to understand and decipher.

ITEM NO.	LDC FULL SERVICE BILLING		
1	Billing Period	07-14 to 08-14	
2	Rate	2 – General Service	
3	Meter Number	476J2	
4	Current Actual	507041	
5	Previous Actual	506180	
6	Difference	861	
7	Meter Constant	(2.5) (861 x 2.5 = 2153 Ccf)	
8	Pressure Correction Factor	1.120	
9	Actual Usage	2411.4 Ccf (2153 x 1.120)	
10	Therm Conversion	1.029 Btu factor	
11	Total Therms	2481.3 Therms (2411.4 x 1.029)	
	BILLING ITEMS	SUBTOTALS	ACCUM. TOTALS
12	Purchased Gas Adjustment (2481.3 x $0.04123)	$102.30	$102.30
13	Customer Charge	$82.15	$184.43
14	Delivery Charge First 1000 therms ($0.32361 x 1000) Over 1000 therms ($0.13682 x 1421.3)	$423.61 $202.67	$810.73
15	Environmental Charge ($0.00413 x 2481.3 therms)	$10.25	$820.98
16	Natural Gas Charge ($0.67912 x 2481.3 therms)	$1,685.10	$2,506.08
17	State Tax (5.7% x $2,506.08)	$142.85	$2,648.93
18	Natural Gas Revenue Tax ($0.024 x 2481.3 therms)	$59.55	$2,708.48
19	Previous Balance	$2,271.86	—
20	Payment	– $2,271.86	—
21	TOTAL CURRENT CHARGE		$2,708.48
22	Cost per Therm	($2,708.48 ÷ 2481.3)	$1.092
23	Cost per Dekatherm	($2,708.48 ÷ 248.1)	$10.92

Figure 14-1. LDC Billing Example

EXPLANATION OF FIGURE 14-1

Composite Natural Gas Billing Example
(LDC Full Service Billing)

Item #1—Billing Period (07-14 to 08-14)

Shown here are the *actual days* period included in this billing—July 01 to August 01—(31) days. The days in individual billings will vary generally from (28) days to (31) days, depending upon the actual billing month. This may seem of little consequence, but it will be understood as this billing is analyzed that there are items on the billing that vary based upon the days in the billing.

Item #2—Rate (2—General Service)

The rate classification is the LDC rate schedule identification that applies to this usage characteristic for the meter being utilized. The complete description of this rate classification can be obtained from the LDC. The actual classification rate nomenclature (2) mayor may not have any particular meaning. This particular example is a General Service Rate #2. Included in this rate are the natural gas commodity and the required delivery services.

Item #3—Meter Number (476J2)

This is the customer's individual meter number. The number shown here should match the number on the physical customer meter. If the billing meter number and the physical meter number do not match, there is a problem. This discrepancy will cause an error in the natural gas billing since the meter number has a direct correlation to Item #7—Meter Constant. Always investigate the accuracy of the billing meter number since it is imperative to an accurate natural gas cost.

Item #4—Current Actual (507041)

This reading is the actual meter reading recorded at the end of the current filling period (08-14).

Item #5—Previous Actual (506180)

This reading is the actual meter reading recorded at the end of the previous billing period. This reading was the previous billing period ending reading. The difference between this reading and the reading in Item #4 gives the figure shown in Item #6.

Item #6—Difference (861)

This figures is the result of subtracting the figure in Item #5 (504889) from the figure in Item #4 (507041). At this point, the (861) number has no specific relevance. It is simply the raw meter data.

Item #7—Meter Constant (1)

This meter has a constant of (2.5), which means that the actual meter reading data must be multiplied by (2.5) to total the true flow of natural gas through the meter for the period from the beginning to the ending reading. Meters can have constants or multiplying factors from (1) to (1,000) or greater.

The constant is based upon relationship of the natural gas flow through the meter to the actual total flow of natural gas. Typically, a large flow has a larger constant, and a smaller flow, a smaller constant. A larger meter constant is utilized to allow a relatively large total natural gas flow through a relatively small meter.

The constant figure is calculated based upon actual natural gas flow compared to meter readings over a specific time period. Neither a large or small constant is more accurate if the meter is correctly calibrated.

Item #8—Pressure Correction Factor (1.120)

This figure is a multiplying factor to calculate natural gas flow based upon pressure differentials. Since natural gas is compressible, the greater the pressure, the larger will be the flow in a given time period.

Standard line pressures are calculated and approved for all natural gas LDCs. When actual pressures vary from the norm, correction factors are calculated and utilized to adjust actual natural gas flow to the standard required by a specific LDC.

Item #9—Actual Usage (2411.4 Ccf)

Here shown is the actual usage in Ccf for the current billing period. This total is calculated as follows:

Item #4	507041	Current actual
Item #5	– 506180	Previous actual
Item #6	861	
Item #7	× (2.5)	Meter constant
	2153	Ccf
Item #8	× 1.120	Pressure correction factor
	2411.4	Ccf

Item #10—Therm Conversion (1.029 Btu factor)

This calculation is utilized to change the volumetric measurement (Ccf) into a thermal value (therm—100,000 Btu). All natural gas flows volumetrically, but it is measured thermally from its wellhead origination point to the customer's serving natural gas LDC.

When the LDC measures in volumetric units, a conversion is made from thermal to volumetric units. This conversion represents the true thermal (heat/Btu) value of the natural gas that the customer is receiving. Typically. (1) Ccf of natural gas contains 103.000 Btu or 1.03 therms.

In this particular billing period, the actual average Ccf heat value was 1.029 or 102,900 Btu per Ccf of natural gas.

Item #11—Total Therms (2481.3 Therms)

Here is shown the actual usage in therms for the current billing period. This total therm usage equals 248.130.000 Btu (2481.3 × 100,000) for the entire billing period. This total is calculated as follows:

Item #9	2411.4	Ccf
Item #10	× 1.029	Therm conversion
Item #11	2481.3	Therms

Item #12—Purchased Natural Gas Adjustment (PGA) ($102.30)

This cost adjustment represents the LDC's additional cost for natural gas above what is included in the natural gas charge entry—Item #16. LDCs utilize the PGA charge when they have fixed natural gas cost in a rate filing for some specific time period, generally 3, 6, 9, or 12 months.

If during one of these fixed cost periods the actual natural gas cost is either greater or less that the fixed natural gas cost, compensating adjustments can be made through the PGA. This adjustment can be an additional cost or credit depending upon the relationship of the actual versus fixed natural gas costs. These costs, whether plus or minus, are straight pass-throughs with no LDC margin added.

Item #13—Customer Charge ($82.15)

The customer charge is the basic LDC charge to obtain service under the applicable rate. Customer charges are detailed in the appropriate rate schedule. Customer charges range from small to very large, depending upon the type and size of the customer's usage characteristics.

In this particular example, the customer charge is based upon a daily charge, which when multiplied by the number of days in the billing period, gives the total billing period customer charge. In this example, the daily charge is $2.65. This charge times the days in the billing period (31) gives the total customer charge of $82.15. A shorter billing period would result in a lower total charge while a longer period would result in a larger total charge.

LDCs seem to be utilizing the daily customer charge concept more frequently than was the case in the past. From an LDC's perspective, the daily charge would seem to make sense since the services occur on a daily basis.

Item #14—Delivery Charge ($423.61 & $202.67)

This charge is what is called a step rate. There are two rates applicable to the units used:

Step #1	First 1000 therms
Step #2	Over 1000 therms

In this type of step rate, there is a large difference between Step #1 and Step #2 costs.

Item #15—Environmental Charge ($10.25)

This charge is a state mandated fee on all natural gas users (residential, commercial, and industrial) for the purpose of cleaning up the environment. What part the usage of natural gas has on cleaning up the environment is unclear but the charge is on all natural gas used by any customer class. This charge is on all natural gas used, whether purchased through the LDC or by a customer from a marketer.

Item #16 —Natural Gas Charge ($1,685.10)

This charge represents the actual gas commodity cost. The natural gas charge and the purchased gas adjustment cost ($102.30)—Item #12—are the total natural gas costs on this billing. Out of a total current charge of ($2,708.48)—Item #21, the natural gas commodity cost is ($1,787.40)—Item #12 plus Item #16—or 66% of the billing total.

Item #17—State Tax ($142.85)

This is a statewide tax and it applies equally to all LDC customers unless there are specific sales tax exemptions that can apply to a specific customer. To determine whether exemptions exist and if you qualify for an exemption, you need to contact the state department of taxation or revenue. Do not contact the LDC since they cannot help you in this matter. If there are state sales tax exemptions, and if you qualify, the state will issue you an exemption number that you will submit to the appropriate LDC to have the tax reduced/removed from the monthly billing.

An important feature of most state sales tax exemption provisions is that a customer can get back state sales tax funds paid in the past if the customer legally was not required to pay the tax. The refund period, based upon specific state statute of limitation provisions, generally are from 3-5 years in the past.

Item #18—Natural Gas Revenue Tax ($59.55)

This charge is assessed by the state for the purpose set forth in the applicable state statute. Typically, charges of this type are for humanitarian purposes; e.g., assistance to low income residential natural gas customers who cannot pay their natural gas billings, especially during the winter months.

Item #19—Previous Balance ($2,271.86)

This figure represents the previous period billing amount. Next billing periods previous balance will be ($2,708.48)—Item #21, this billing period's total current charge.

Item #20—Payment ($2,271.86)

This figure represents the customer payment received by the LDC for the previous billing period. In this example, the customer paid the total due for the previous billing period. If total payment had not been received by the LDC, the total due amount would be added to this period's total current charge, together with any late charge penalties.

Item #21—Total Current Charge ($2,708.48)

This amount includes Item #'s 12, 13, 14, 15, 16, 17, and 18. If there is not a previous balance due, this amount represents the total amount the customer currently owes the LDC for the current billing period.

Item #22—Cost Per Therm ($1.092)

This item is not an LDC billing entry. It is shown here to detail the actual customer burner tip (meter point) cost per therm (100,000 Btu).

Item #23—Costs Per Dekatherm ($10.92)

This item, like Item #22, is included to detail the actual cost per dekatherm (1,000,000 Btu).

ITEM NO.	LDC DELIVERY SERVICE							
1→	NATURAL GAS BILLING 01-01 through 02-01 (31 DAYS)							
	2 ↓	3 ↓	4 ↓	5 ↓	6 ↓	7 ↓	8 ↓	9 ↓
	Meter Number	Begin Reading	End Reading	Difference	Meter Constant	Therm Factor	Volumetric (Ccf) Factor	Total Therm Usage
	0721632	49320.1	56710.5	7390.4	10	1.000	1.029	76047.6

ITEM NO.	TARIFF RATE LG – FIRM DELIVERY SERVICE		
10→	BILLING DETAIL	SUBTOTALS	ACCUM TOTALS
11→	**Customer Charge**	**$271.56**	$271.56
12→	**System Delivery Charge** (58,117.2 therms @ $0.061273)	**$3,561.02**	$3,832.58
13→	**Natural Gas Environmental Charge** (58,117.2 therms @ $0.007128)	**$414.26**	$4,246.84
14→	**Excess Natural Gas Charge** (17,930.4 therms @ $0.031782)	**$569.86**	$4,816.70
15→	**Unauthorized System Natural Gas Charge** (17,930.4 therms @ $0.831200)	**$14,903.75**	$19,720.45
16→	**Unauthorized System Delivery Charge** (17,930.4 therms @ $0.072182)	**$1,294.25**	$21,014.70
17→	**Unauthorized Natural Gas Penalty** (17,930.4 therms @ $0.106213)	**$1,904.44**	$22,919.14
18→	**Bank Imbalance Charge** (17,930.4 therms @ $0.201667)	**$3,615.97**	$26,535.11
19→	**Administrative Charge**	**$150.00**	$26,685.11
20→	**Facility Charge**	**$110.00**	$26,795.11
21→	**Total Billing Amount**		$26,795.11
22→	**State Natural Gas Revenue Charge —** @ 3.5%	**$937.83**	$27,732.94
23→	**State Regulatory Agency Charge —** @1.6%	**$428.72**	$28,161.66
24→	**State Sales Tax —** @ 7.1%	**$1,902.45**	$30,064.11
25→	**Current Amount due**		$30,064.11
26→	**Prior Amount Due**	**$7,116.03**	$37,180.14
27→	**Prior Amount Due Charge —** @ 1.5%	**$106.74**	$37,286.88
28→	**TOTAL AMOUNT DUE**		**$37,286.88**
29→	**Banked Natural Gas Balance 01-01**	**7620 therms**	
30→	**Prior Period Adjustments (+)**	**210 therms**	
31→	**Current Period Adjustments (–)**	**25,760.4 therms**	
32→	**Banked Natural Gas Balance 02-01**	**(0)**	
33→	**Bank Limit**	**9167 therms**	
34→	**Peak Usage Month**	**October**	

Figure 14-2. LDC Delivery Service Billing Example

EXPLANATION OF FIGURE 14-2

Composite Natural Gas Billing Example
(LDC Delivery Service)

Item #1—Billing Period (01-01 through 02-01)

Shown here are the actual days period included in this billing—January 01 through February 01—(31) days. The days in individual billings will vary generally from (28) days to (31) days depending upon the actual billing month. This may seem of little consequence, but it will be understood as this billing is analyzed that there are items on the billing that vary, based upon the days in the billing.

Item #2—Meter Number (0721632)

This section shows the customer's individual meter number together with the beginning and ending meter readings for natural gas usage. The number shown in this section of the billing should match the number on the physical customer meter.

If the billing meter number and the physical meter number do not match, there is a problem. This discrepancy will cause an error in the natural gas billing since the meter number has a direct correlation to Item #6—Meter Constant. Always investigate the accuracy of the billing meter number since it is imperative to an accurate natural gas cost.

Item #3—Begin Reading (50081)

The beginning figure indicates the beginning numeric meter reading when the previous month's meter reading was taken. The previous month's ending reading and this month's beginning reading should be the same reading.

Think of the meter reading as similar to a nonresetable automobile odometer reading. When a replacement meter is installed in a customer location, its recording register will be set to (0). The entire time that the meter is in service at the customer's location, this register will continue to read on an accumulating basis. An actual month's natural gas usage will be calculated by subtracting last month's ending reading (the billing month's beginning reading) from the billing month's ending reading.

Item #4	56710.5
Item #3	– 49320.1
Item #5	7390.4

Remember, this total represents only the raw meter reading data. The reading total (7390.4) Item #5—Difference, when multiplied by (10) Item #6—Meter Constant, is utilized as a factor to arrive at the actual month's natural gas.

Item #4—End Reading (56710.5)

The end figure indicates the actual meter reading at the end of the billing period—in this example—(02-01).

Item #5—Difference (7390.4)

The meter register beginning reading is subtracted from the ending meter register reading. (See Item #3) The meter register reading represents only the raw meter data—it does not represent actual usage for the billing period.

Item #6—Meter Constant (10)

Natural gas meters typically do not measure the entire natural gas flow that passes through. Rather, on a continuous basis, meters measure on a ratio basis that is indicated in this case by the (10)—Meter Constant. The meter-multiplying constant is a meter calculation, determined by the actual ratio of flow to meter recording.

Item #5	7390.4	difference
Item #6	× 10	meter constant
	73904	

Meter constants can range from (1) to (1,000) or greater. If the meter constant is correct, a small meter constant is no more accurate than a large meter constant.

Generally, meter accuracy is calibrated to at least ±1/10% (.01) over the full meter recording range. As a rule, natural gas meters do not measure total flow, but rather measure at a ratio to total actual flow. Smaller, less expensive meters can be utilized by incorporating a constant factor that allows the meter to actually measure only a portion of actual natural gas flow.

Item #7—Therm Factor (1.000)

All natural gas is purchased at wellhead and transported on an interstate basis on a therm (Btu) basis. Thermal value of natural gas is not directly related to volumetric values. In Item #7, the thermal value is show as (1.000) or (1) therm of natural gas contains (100,000 Btu).

If the thermal value were different, it would be entered here. This item is a multiplying factor, but since it has the value of (1) it does not change any value.

Item #8—Volumetric (Ccf) Factor (1.029)

Natural gas, from the wellhead meter point to the LDC city gate receipt point, is always measured thermally in (therms and dekatherms). Once natural gas is delivered to the serving LDC's city gate, it may be measured in volumetric united (Ccf or Mcf). Thermal measurements are absolute (always the same).

When the measurement criterion is changed to volumetric units, a multiplying factor may be required. In this example, the factor is (1.029) or, (1) Ccf of natural gas equals (1.029) therms of natural gas that was delivered to the LDC city gate receipt point. The reason for the difference between thermal and volumetric measurements is that all natural gas has varying (Btu) values.

For the month of this billing, the average heat value of the natural gas delivered was (1.029 Btu) for each Ccf of natural gas used.

Item #5 × #6 × #7	73904	meter constant value total
Item #8	× 1.029	volumetric (Ccf) factor
	76047.6	therm usage

Item #9—Total Therm Usage (76047.6)

Here is shown the total therms consumed during the billing period. This total is calculated as follows:

Item #5	7390.4	difference
Item #6	× 10	meter constant
	73904	meter constant value total
Item #7	× 1.000	therm factor
	73904	meter constant value total
Item #8	× 1.029	volumetric (Ccf) factor
	76047.6	total therm usage

Item #10—Rate (LG—Firm Delivery Service)

The rate classification is the LDC's rate schedule identification that applies to the usage characteristics for the meter being utilized. The complete description of this rate classification can be obtained from the LDC. In this example, the rate classification is (LG). The actual classification nomenclature (LG) may or may not have any particular meaning depending upon the specific LDC involved; however, in this particular example, (LG) means large general service.

Also, in this specific instance, this rate is for firm delivery service only. In addition to this billing, there will be a second billing from the natural gas marketer. (See Figure 14-3)

Item #11—Customer Charge ($271.56)

The customer charge is the basic LDC charge to obtain service under the applicable rate. Customer charges are detailed in the appropriate rate schedule. Customer charges range from small to very large, depending upon the type and size of the customer's usage characteristics.

In this particular example, the customer charge is based upon a daily charge, which when multiplied by the number of days in the billing period, gives the total billing period customer charge—in this example the daily charge is $8.76.

Item #11	$ 8.76	daily charge
Item #1	× 31	days
	$ 271.56	total customer charge

A shorter billing period would result in a lower total charge, while a longer period would result in a larger total charge. LDCs seem to be utilizing the daily customer charge concept more frequently than was the case in the past. From an LDC's perspective, the daily charge would seem to make sense, since the services provided are on a daily, not monthly, basis.

Item #12—System Delivery Charge ($3,561.02)

The system delivery charge is to transport a customer's natural gas through the LDC's intrastate pipeline distribution system. This charge includes the cost of maintaining the LDC pipeline, rental of the pipeline, metering of customer natural gas quantities, natural gas shrink (line loss), and miscellaneous fees and charges. The complete listing of items included in this charge is itemized in the complete rate schedule.

Item #13—Natural Gas Environmental Charge ($414.26)

The natural gas environmental charge is a state mandated fee on all natural gas users (residential, commercial, industrial) for the purpose of cleaning up the environment. What part the usage of natural gas has on cleaning up the environment is unclear, but the charge is on all natural gas used by any customer class. This charge is on all natural gas used, whether purchased through the LDC or by a customer from a marketer.

Item #14- Excess Natural Gas Charge ($569.86)

The excess natural gas charge, together with Item #17, constitutes what this LDC charges a customer that is transporting their own natural gas when that customer uses more natural gas than their marketer sent to the LDC city gate receipt point during the billing period.

This excess natural gas charge (17,930.4 therms) is calculated by taking total therm usage (Item #9: 76,047.6 therms), minus marketer-provided and banked natural gas (Item #12: 58,117.2 therms), times the incremental cost of ($0.031782)—(76,047.6 therms- 58,117.2 therms × $0.031782 = $569.86)

Item #15—Unauthorized System Natural Gas Charge ($14,903.75)

The unauthorized system natural gas charge is (17,930.4 therms × $0.831200 = $14,903.75). This charge is what the LDC charges for customer natural gas replacement during non-/under-delivery periods by the customer's marketer. The actual penalty cost of this item would be the difference the natural gas cost here as compared to the natural gas cost if provided by the customer's marketer. The marketer generally should reimburse this customer for this cost.

Item #16—Unauthorized System Delivery Charge ($1,294.25)

The unauthorized system delivery charge is (17,930.4 therms × $0.072182 = $1,294.25). This item and Item #12—System Delivery Charge are for the same thing moving natural gas through the LDC intrastate pipeline distribution system. The penalty here is the difference between the incremental cost in Item #12 and the cost used in this item.

Why the cost difference between Items #12 and #16? In Item #12, the natural gas transported was provided by the customer. In Item #16, the natural gas was provided by the LDC even though the customer's marketer was responsible for providing the natural gas. Generally, the marketer should reimburse the customer for this cost.

Item #17—Unauthorized Natural Penalty ($1,904.44)

The unauthorized natural gas penalty is (17,930.4 therms × $0.106213

= $1,904.44). This is simply a penalty to the customer for their marketer's failure to deliver the natural gas required to cover the customer's actual usage during the billing period. This type of penalty is typically what an LDC classifies as their cost to acquire natural gas that the customer used but had not requested the LDC to provide. It could be considered an inconvenience cost to the LDC to provide natural gas that they had not been contracted to provide. The marketer generally should reimburse the customer for this cost.

Item #18—Bank Imbalance Charge ($3,615.97)

The bank imbalance charge is (17,930.4 therms × $0.201667 = $3,615.97). This charge is related to the natural gas bank in this billing period being completely depleted, requiring the LDC to provide system-supply natural gas to a customer that has contracted with this LDC to provide only intrastate transportation of customer-provided natural gas. The marketer generally should reimburse the customer for this cost.

Item #19—Administrative Charge ($150.00)

The administrative charge is a charge that most LDCs include to be compensated for the paper work involved in tracking and recording the various items required in the customer/marketer transportation process.

Item #20—Facility Charge ($110.00)

The facility charge, when present on a natural gas billing, generally indicates a non-standard condition. Usually this type of charge reflects equipment purchases and/or leases by the customer from the LDC. In the right situation, this type of arrangement can benefit a customer.

The point that has to be considered by the customer is do these charges ever end? Generally, they do not. The first thing to do is:

1) Locate a copy of the facility charge and/or lease written agreement (contract)
2) Determine whether the item is required
3) Determine what the buyout value is for the equipment charge and/or lease

If you cannot find your contract copy, contact the LDC and ask for a copy of the particular agreement. For example, let's assume the value of the item is as follows:

1) Facility charge equipment—buyout value $2,000

2) The facility charge equipment could be purchased for less than (19) billing period payments—($2,000 + $110 = 18.2 months)

Assuming the equipment still has value to the facility and the customer can maintain it, or the equipment can be contracted for a third-party maintenance agreement, then serious consideration should be given to a buyout of the equipment.

Item #21—Total Billing Amount ($26,795.11)

This amount is a total of all Items #11-#20. This amount does not represent the actual amount the customer will pay to the LDC. To this figure must be added the amounts in Item #'s 22, 23, and 24.

Item #22—State Natural Gas Revenue Charge ($937.83)

The state natural gas revenue charge is assessed by the state for the purpose set forth in the applicable state statute. Typically, charges of this type are for humanitarian purposes—e.g. assistance to low income residential natural gas customers that cannot pay for natural gas billings, especially during winter months.

Item #23—State Regulatory Agency Charge ($428.72)

The state regulatory agency charge is imposed by the state regulatory agency. Its purpose is to provide revenue for the regulatory commission for either its normal or some special activity.

Some regulatory agencies receive state financing and some have to fund themselves through individual LDC billing charges. Whatever the reasons for this charge, the customer has to pay it as it cannot be negotiated away. The particulars of a charge of this type can be determined by contacting the specific regulatory agency involved.

Item #24— State Sales Tax ($1,902.45)

The state sales tax is a statewide tax and applies equally to all LDC customers unless there are specific sales tax exemptions that can apply to a specific customer. To determine whether exemptions exist and if a customer qualifies for an exemption, contact the state department of taxation or revenue. Do not contact the LDC since they cannot help in this matter. If there are state sales tax exemptions and the qualifications are there, the state will issue an exemption number that can be submitted to the appropriate LDC to have the tax reduced or removed from the customer's monthly billing.

An important feature of most state sales tax exemption provisions is that a customer can get back state sales tax funds paid in the past if the customer legally was not required to pay the tax. The refund period, based upon specific state statute of limitation provisions, generally are from 3-5 years in the past.

If a customer is paying state sales tax and has a legal right not to pay, it is the customer's responsibility to contact the state and take the appropriate action to have the tax reduced or eliminated. It is not the state's responsibility, and certainly not the LDC's responsibility to perform this procedure. The state assumes all LDC customers owe the tax—it is the customer's responsibility to prove otherwise. The particulars of a charge of this type and the potential for exemption can be determined by contacting the specific state imposing the sales tax.

Item #25—Current Amount Due ($30,064.11)

This is the total amount due for all Item #'s 11-24. If there are no past due accounts unpaid, this current amount due will be what the customer owes to the LDC. If there are past account payments due, the charges will be added to the amounts shown in this item. (See Items #26 and #27.)

Item #26—Prior Amount Due ($7,116.03)

Here is shown prior amounts not yet paid. This amount can be for the billing period immediately prior to this billing period, or it can be an accumulation of several past billing periods.

Item #27—Prior Amount Due Charge ($106.74)

This amount is for the interest charged by the LDC to the customer for the prior amount due (Item #26).

Item #28—Total Amount Due ($37,286.88)

The amount shown is the total amount the customer owes the LDC including the current billing charge (Item #21), plus the past due amount from previous billing periods (Items #26 and #27).

Item #29— Banked Natural Gas Balance 0101 (7620 therms)

The banked natural gas balance indicates the amount of natural gas the customer had in storage (bank) in the LDC system at the beginning of the current billing period. This amount represents natural gas acquired for the customer by the marketer that was in excess of what was actually used by the customer. A balance of natural gas is not a bad thing to have since it is virtually impossibly for the marketer to deliver the exact amount of natural gas that will be used by the customer in a given time period.

In recognition of this, most, if not all LDCs have customer banking (storage) of natural gas rate provisions. These provisions allow the customer to carry some maximum amount of natural gas in the LDC system to cover shortfalls of actual gas deliveries compared to actual gas usage.

It is generally a good thing to utilize customer natural gas storage capability to the full extent possible since it can reduce and/or eliminate the possibility of LDC penalties on the natural gas used in excess of marketer deliveries to LDC city gate receipt point. Depending upon the specific LDC, there may be injection, storage, and extraction fees associated with utilization of the banking capability.

Item #30— Prior Period Adjustments (210 THERMS)

The prior period adjustments figure represents an adjustment to the banked natural gas quantity that occurred after the banked balance date on 01-01. An adjustment of this type is common since final meter readings at the LDC city gate and customer's burner tip are often trued up after the next billing period begins.

Item #31— Current Period Adjustments (25,760.4 THERMS)

The current period adjustments figure represents customer natural gas usage in excess of marketer deliveries in the current billing period. This figure is calculated as follows:

Item #29	7,620.0	bank on 01-01
Item #30	+ 210.0	adjustment
	7,830.0	total bank
Item #14	+ 17,930.4	excess natural gas use
	25,760.4	total excess natural gas use

In the current billing period, the total natural gas usage is as follows:

50,287.2	marketer natural gas delivered
17,930.4	excess natural gas delivered
7,620.0	bank on 01-01
210.0	adjustment
76,047.6	total natural gas delivered

In this particular billing period, natural gas usage exceeded gas deliveries by 25,760.4 units, And after depleting the bank of 7,830 units, the natural gas usage was still 17,930.4 units more than what was available. These 17,930.4 units were the unauthorized units in Items #15, #16, and #17.

Due to the imbalance between deliveries, bank quantities, and actual usage, there are penalties for the excess natural gas quantities. Generally the marketer should reimburse the customer for these penalty costs.

Item #32—Banked Natural Gas Balance 0201 (0)

Because of the under delivery and over use of natural gas during this billing period, there is no banked natural gas available to help cushion the next billing period variations. The marketer should be watching the bank status and adjust deliveries to maximize bank potentials.

Item #33— Bank Limit (9,167 therms)

The bank limit figure is the total bank limit that is allowed in any billing period. It is generally calculated on a customer-specific basis and on a rolling (11) month ratchet basis.

This means that the highest natural gas usage month in the last (11) months is used to determine the maximum bank quantity. Generally, the highest or peak month natural gas usage bank quantity is allowed to be 10% of actual usage for that billing period. In this example, the peak month usage was 91,670 units or a bank maximum quantity allowable of 9,167 units (10% of 91,670).

Item #34— Peak Usage Month (October)

Here is shown the peak usage month in the last (11) billing months. If next billing period usage quantity exceeds the quantity used last October, the maximum bank quantity will be adjusted accordingly. If there are no monthly usages greater than last October's usage by the end of next September, the next highest monthly usage in the last (11) months will become the basis for banked balance maximum quantities.

Since this billing example is for LDC delivery services only, the marketer-provided natural gas billing also needs to be understood and evaluated.

ITEM NO.	MARKETER NATURAL GAS DELIVERY SERVICE		
1→	**Natural Gas Commodity Billing Period**		**01-01 though 02-01**
2→	**Customer Contract Number**		**#14-A/B (FIRM)**
	BILLING DETAIL	SUBTOTALS	ACCUM. TOTALS
3→	**Wellhead Natural Gas Cost** (21,610.6 therms @ $0.5712))	**$12,343.97**	$12,343.97
4→	**Wellhead Natural Gas Cost** (14,186.9 therms @ $0.4719)	**$6,694.80**	$19,038.77
5→	**Wellhead Natural Gas Cost** (14,489.7 therms @ $0.6123)	**$8,872.04**	$27,910.81
6→	**Pipeline Cost - Firm** (22,974.8 therms @ $0.0261)	**$599.64**	$28,510.45
7→	**Pipeline Cost - Firm** (27,312.4 therms @ $0.0296)	**$808.45**	$29,318.90
8→	**Pipeline Retention** (50,287.2 therms @ $0.0021)	**$105.60**	$29,424.50
9→	**Total Natural Gas Cost**		$29,424.50
10→	**State Natural Gas Tax** (@ 3.5%)	**$1,029.85**	$30,454.35
11→	**Current Amount Due**		$30,454.35
12→	**Past Billing Amount**	$28,710.06	—
13→	**Past Due Amount**	-0-	—
14→	**TOTAL AMOUNT DUE**		**$30,454.35**
15→	Cost per Therm	($30,454.35 ÷ 50,287.2)	$0.6056
16→	Cost per Dekatherm	($30,454.35 ÷ 5,028.72)	$6.0560

Figure 14-3. Composite Natural Gas Billing Example (Marketer Natural Gas Delivery Service)

EXPLANATION OF FIGURE 14-3

Composite Natural Gas Billing Example
(Marketer Natural Gas Delivery Service)

Item #1—Natural Gas Commodity Billing Period (01-01 through 02-01)

This billing period is for (31) days (01-01 through 02-01). This period should be the same as for the LDC billing

Item #2—Customer Contract Number (#14-A/B—Firm)

Here is shown the customer/marketer specific contract identification data. This information shows that the contract is (#14) and has either two sections (A/B), or has two revisions (A/B) since the original inception of the contract (#14),

Also, shown is the fact that this is a firm service contract. This means that both the natural gas commodity and all interstate transportation to the LDC city gate receipt point is Firm (non-interruptible), except for force majeure conditions. This class of service should match the LDC delivery service class.

Item #3— Wellhead Natural Gas Cost ($12,343.97)

The wellhead natural gas cost is calculated and shown as part of the natural gas cost for the current billing period. Notice that not all of the natural gas purchased was at the same price. This natural gas was purchased at ($0.5712 per therm), while gas purchased in Items #4—($0.4719 per therm), and Item #5—($0.6123 per therm).

Why the difference? It appears that for some reason, probably marketer capacity related, that three entirely separate natural gas purchases were made, all at different prices. If all natural gas had been purchased at the Item #4 price ($0.4719), the wellhead natural gas cost total would have been $4, 180.28 less than it was in actuality.

In addition to this cost differential, (Figure 14-2, Items #15, 16, and 17) there were penalties for unauthorized LDC natural gas usage under deliveries by the marketer. Furthermore, to all of this, the marketer did not balance the customer bank natural gas quantities.

In Figure 14-2, Item #18, there is an imbalance charge (penalty) created because of the marketer's non-delivery of bank natural gas. Also, in Figure 14-2, Item #32, the banked natural gas balance at the end of this billing period is (0). This means that if the marketer does not deliver at least the actual quantity of natural gas used by the customer in the next billing period, there will be further penalties to the customer from the LDC.

Item #4—Wellhead Natural Gas Cost ($6,694.80)

See Item #3 for a discussion of this same type of item.

Item #5— Wellhead Natural Gas Cost ($8,872.04)

See Item #3 for a discussion of this same type of item.

Item #6—*Pipeline Cost—Firm* ($59.64)

The pipeline cost item, as with wellhead natural gas sost (Items #3, #4, and #5), there are separate cost figures for the interstate pipeline transportation. The reason for the different pipeline costs are not explained on this billing, but the reason is probably related to the marketer's requirements during this billing period as compared to what was scheduled and/or ordered.

Item #7— Pipeline Cost—Firm ($808.45)

See Item #6 for a discussion of this same type of item.

Item #8—Pipeline Retention ($105.60)

The pipeline retention is calculating the cost by the transporting interstate pipeline for natural gas losses. These losses generally have at least two distinct causes.

1) All pipelines leak natural gas
2) Usage of pipeline natural gas supply for upstream pressure boost station operation

Generally, these losses together total 3-8% of the total pipeline throughput. All customers who utilize the pipeline will be charged the retention cost. All pipelines have different retention factors and/or costs depending upon pipe age, operation pressures, distance of pipeline, etc. All retention costs are presented to and approved by FERC.

Item #9—Total Natural Gas Cost ($24,424.50)

This amount includes Items #3, #4, #5, #6, #7, and #8. The cost shown here represents the total marketer charge to the customer for natural gas, transportation, and all related services as detailed in Item #2—Customer/Marketer Contract.

Item #10— State Natural Gas Tax ($1,029.85)

The state natural gas tax charge is assessed by the state for the purpose set forth in the applicable state statute. Typically, charges of this type are for humanitarian purposes; e.g., assistance to low income residential natural gas customer that cannot pay for natural gas billings, especially during winter months.

Item #11—Current Amount Due ($30,454.35)

This amount includes Items #3, #4, #5, #6, #7, #8, and #10. Excluding previous amounts yet unpaid, the current amount due is the total due to the marketer by the customer for the current billing period.

Item #12—Past Billing Amount ($28,710.06)

Here is shown the amount of the past billing period billed to the customer. If this amount remains unpaid at the time of this billing, it will be shown in Item #13.

Item #13—Past Due Amount (-0-)

If the customer had not paid for previous billing period purchases, the past due amount would be shown here and added to this billing period's amount due (Item #11).

Item #14—Total Amount Due ($30,454.35)

This amount includes both the current amount due (Item #11) and any past amounts due (Item #12). This is the total amount due to the marketer from the customer for this billing period.

Item #15—Cost Per Therm ($0.6056)

This item is not a marketer billing entry. It is shown here to detail the actual customer LDC city gas natural gas cost per therm (100,000 Btu).

Item #16—Cost Per Dekatherm ($6.0560)

This item, like Item #15, is included to detail the actual cost per dekatherm (1,000,000 Btu).

Total Natural Gas Cost for Figures 14-2 and 14-3

Figures 14-2 and 14-3 are shown opposite.

Figure 14-2.
Composite Natural Gas Billing Example Delivery Service Only

Natural Gas Used

Item #12	58,117.2	therms
Item #14	17,930.4	therms
TOTAL NATURAL GAS USED	76,047.6	THERMS

Natural Gas Transportation Cost

Item #25	$30,064.11

Natural Gas Transportation Cost Per Therm

($30,064.11 + 76,047.6) = $0.3953 Per Therm

Figure 14-3.
Composite Marketer Natural Gas Billing Example

Natural Gas Used

Item #8	50,287.2 THERMS

Natural Gas Cost

Item #11	$30,454.35

Natural Gas Transportation Cost Per Therm

($30,454.35 + 50,287.2) = $0.6056 Per Therm

TOTAL COST OF NATURAL GAS

1.	Figure 14-2 cost per therm	$0.3953
2.	Figure 14-3 cost per therm	$0.6056
TOTAL COST PER THERM		$1.0009
TOTAL COST PER DEKATHERM		$10.009

Excess Costs in Figure 14-2

Item #	*Item*	*Amount*
14	Excess natural gas charge	$ 569.86
15	Unauthorized system natural gas charge	$ 14,903.75
16	Unauthorized system delivery charge	$ 1,294.25
17	Unauthorized natural gas penalty	$ 1,904.44
18	Bank imbalance charge	$ 3,615.97
27	Prior amount due charge	$ 106.74

Illustrated here are various items that are related to marketer non-delivery penalties, and the customer penalty for non-payment of previous billings. All of the Items #14, #15, #16, #17, #18, and #27 are bottom-line penalties.

In Item #15, the penalty is the difference between what the LDC charged per therm ($0.8312) and what the customer would have paid to the marketer ($0.5851). ($0.8312 – $0.5851 × 17,930.4 therms = $4,412.67.)

These figures represents an actual customer billings. What went wrong? The customer's inattention to detail! This billing demonstrates what can happen when a customer relies on either a marketer or LDC to take care of them.

Unexplained Cost Variations in Figure 14-3

Item #	*Item*	*Amount*
3	Wellhead natural gas cost	$ 0.5712
4	Wellhead natural gas cost	$ 0.4719
5	Wellhead natural gas cost	$ 0.6123
6	Pipeline cost	$ 0.0261
7	Pipeline cost	$ 0.0296

In this marketer billing, there are shown (3) separate natural gas purchases at costs ranging from $0.4719-$0.6123 per therm. Also, there are (2) separate pipeline costs, $0.0261 and $0.0296 per therm.

There are occasions where a marketer will be required to acquire natural gas at varying costs, especially if a customer changes their usage drastically during a specific billing period.

The information in Figures 14-2 and 14-3 illustrate what can happen when a good strategy is not tracked on a continuous basis. The value of utilizing a marketer to supply natural gas is greatly reduced when the details of the process are ignored. This chapter has covered actual billing examples of the three different processes: 1) LDC full service, 2) LDC delivery service, and 3) marketer commodity service.

It is very important to understand how all of the various pieces fit together. The bottom line is: *The customer is responsible for their natural gas cost.*

SYNOPSIS—THE NATURAL GAS BILLING PROCESS

Natural gas billing detail is critical to understanding and reducing costs. There may seem to be excess details in the billing process, but all of the various pieces are important to an effective cost control and cost reduction process. Once the process detail is utilized several times, it will be much more understandable.

Chapter 15

The Propane-air System

PROPANE-AIR—HOW IT WORKS

The main components of a typical propane-air backup system are

1. Propane storage tank(s)
2. Fuel delivery station
3. Pressurizing pump
4. Vaporizer
5. Mixer

The output of the propane-air system is normally tied into a customer's natural gas distribution piping after the utility's meter. Propane-air is distributed to fuel-burning equipment through existing natural gas piping without any interruption of supply.

Propane-air—Typical Applications

1. To provide a compatible backup fuel for customers who are on interruptible natural gas service
2. For peak shaving where the customer is either limited in the amount of natural gas available from the LDC, or where the customer purchases natural gas on a transport basis at a fixed rate per day or week
3. To provide all or a portion of a customer's natural gas needs

Propane Properties

Propane vaporizes at –43.8°F at atmospheric pressure. In storage, its vapor pressure is about 23.5 psig at 0° and 175 psig at 100°. Therefore, propane systems must deal with propane both as a gas and a liquid. Pure propane vapor has a specific gravity of 1.52 (heavier than air) compared to air 1.0 and natural gas .60 (lighter than air).

Propane gas will burn only when combined with sufficient oxygen. The flammable range for propane is 2.0% to 9.6% gas in air. For comparison, the flammable range for natural gas is 4.5% to 16% gas in air. Expressed another way, the ideal combustion ratio for propane is 24:1 (96% air, 4% propane) while for natural gas, it is 10:1 (90% air, 10% gas).

Propane contains about 2,520 Btu/scf compared to natural gas with about 1,000 Btu/scf. Liquefied propane contains approximately 92,000 Btu per gallon. Thus, 11.257 gallons of propane contains about 1 MMBtu (or 1 dekatherm) or about as much energy as 1 Mcf of natural gas. While natural gas is difficult to store at a customer site, propane's energy density makes storage reasonably simple.

Identifying a Suitable Application

A potential propane-air application exists where both natural gas and propane are available. What kinds of energy customers might benefit from a propane-air system? Typical small applications include schools, hospitals, nursing homes, warehouses, shopping malls, hotels/conference centers, etc. Larger applications include industrial users, food manufacturers, college campuses, and electrical and/or steam-generating power plants.

Propane System Variables

1. Size of connected gas-consuming load, maximum/minimum flow rates, and maximum consumption per hour and per day.
2. Pressure used for natural gas distribution.
3. Expected changes in gas use (i.e., plant expansion)
4. Insurance requirements (i.e., FM, IRI)
5. Volume of propane storage needed, depending on
   a) Maximum consumption rate and number of days backup desired
   b) Requirements, if any, of natural gas LDC, propane supplier, or other party
   c) Available space and code or insurance constraints
   d) LP gas sources and delivery vehicle size
6. Other uses for propane—such as forklifts or vehicle fleets—should be evaluated to determine propane supply requirements and system hardware.

Codes and Standards

NFPA 58, Standard for the Storage and Handling of Liquefied

Petroleum Gasses (recently renamed the LP-Gas Code) is a common minimum standard referenced in most state codes.

Minimum distances are required between components of the propane system and from buildings and property lines. Good design practice should be observed, which may involve requirements exceeding the minimum requirements of this standard.

Review of a proposed facility by state and local agencies is mandatory. Insurers such as Factory Mutual and Industrial Risk Insurers also have specific requirements. Local code officials and fire departments must be consulted prior to construction and trained in the safety features of any system.

PROPANE-AIR SYSTEM COMPONENTS

Propane Storage Tanks

Propane tanks are available for above-ground or below-ground installation. Underground or mounded storage is increasingly popular, enhancing both aesthetics and fire safety.

Under federal regulations and those in effect in most states, propane is not considered a ground contaminant. Even though double-wall structures are not required for below-grade installations, proper design, installation, and monitoring is essential including complete corrosion-protection systems.

Tanks are available in many sizes. The most commonly used are 1,000, 2,000, 3,900, 12,000, 18,000, and 30,000 gallon gross (liquid) capacity. Net propane capacity is about 85% of total tank capacity. Multiple tanks can be installed to achieve any desired total storage capacity.

Typical tank equipment includes relief valves, excess-flow valves, and gauges for temperature, pressure, and liquid level. A site-specific fire safety system will be required based upon the features of the installed storage system.

Fuel Delivery Station

Most customer backup systems receive propane by truck delivery. A truck unload station provides a convenient and safe connection for the propane supplier to deliver fuel to the tank. Station features include breakaway connections, fill, vent, excess flow, manual isolation, and emergency shutdown valves.

Propane Pump

Because the pressure of propane in a storage tank varies with temperature, many propane-air systems utilize pumps to ensure adequate product pressure feeding the vaporizer. Many varieties of motor-driven pumps are available, including positive displacement and impeller types.

For continuous duty, duplicate pump systems might be required. Optional controls can be applied for auto-start, etc. Pumping a liquid at its boiling point requires careful attention to the pump placement and the design of related piping and controls. Effective vapor elimination is critical.

Propane Vaporizer

Liquid propane is pumped to a vaporizer. This device heats the liquid to the boiling point to produce required volumes of propane vapor. The majority of vaporizers use some of the propane fuel for this task, but electricity, steam, and hot water heat input sources can also be used. Small- to medium-size backup systems often have vaporization equipment packaged with propane-air blending equipment.

Propane-air Blenders (Mixers)

As a general rule, a propane-air mixture has Btu characteristics similar to natural gas. A blender (also called mixer) is used to produce mixed gas in volumes and at a pressure sufficient for distribution in a customer's existing natural gas piping.

Venturi blenders use pressurized propane vapor to move air through a specially designed nozzle. Atmospheric venturi systems are normally used for mixed-gas pressures up to 15 psig.

Very large propane-air flows are best managed by using special blenders to proportion compressed air and propane gas with specific control mechanisms. These special blenders are effective at discharge pressures from less than 10 to 250 psig.

Critical considerations include the accuracy of the device to produce a consistent propane-air mix across the full range of capacity, reliability, serviceability, and safety of operation.

On systems using compressed air, specific gravity or comparable mixed gas monitoring equipment should be integrated in the control system. System operation should provide the customer with a seamless transition to propane-air operation and back to natural gas.

Safety Systems

Beyond minimum code requirements, LP gas safety can be enhanced in several areas at the typical site. Ultimately, local public safety agencies, insurance carriers, the propane system designer, and the owner must define safety systems and emergency response factors appropriate for the site.

Minimizing the potential volume and duration of accidental propane release is an important safety concern. Some protection against a system rupture is provided in basic construction. However, comprehensive emergency shutdown systems that offer improved containment might be required in specific installations.

An emergency shutdown application normally includes actuated valves in all tank openings, fuel transfer connections, and at other locations. Automatic closure of these valves can quickly seal the system. Interlocks in the main system control, site power supply, off-site alarms, etc. can yield safe shutdown and improved emergency response.

Gas and Fire Detection Systems

These can be installed to monitor critical plant areas. Detection can be integrated within an emergency shutdown system for alarm and/or automatic shutdown of the plant.

System Control

As with any process system, control of propane-air backup systems is a flexible concept. A well-designed system will include at least basic integration of the operating equipment and subsystems. However, integrated control of the entire system should be considered to ensure the process is made as safe, simple, and reliable as possible.

Even for smaller systems, advanced electronic process-control technology is easily applied. A well-executed control strategy can deliver improved accuracy and safety, automatic gas quality control, comprehensive monitoring, programmed start/stop routines, and remote operation.

Connection with natural gas measurement equipment can allow automated balancing of natural gas deliveries, avoiding potentially large penalties for using too much, or too little gas.

PROPANE-AIR SYSTEM—PRICING

Propane Data Accumulation

The cost of a propane-air installation varies with its size, capacity, and operating characteristics. Most local distribution companies require the availability of an alternate fuel source for interruptible customer service. A propane-air unit can be a good investment, both from the point of being able to qualify for interruptible natural gas service as well as providing reliable backup fuel supply source.

Following is a listing of practical steps in the process of natural gas cost reduction that outlines the data items needed to calculate the cost of a propane-air backup system.

1. Maximum daily natural gas usage (Mcf/Dth)
2. Days of natural gas replacement desired. (Normally 3-6 day supply is considered sufficient for standby interruptability purposes)
3. Total required quantity of natural gas replacement
4. The liquid propane equivalent requirement for natural gas replacement
5. Propane mixer capacity requirement (including vaporizer, pump and mix unit)
6. Storage tank requirement—standard tank sizes: 1,000 gallon (800-gallon capacity); 18,000 gallon (14,400-gallon capacity); 30,000 gallon (24,000-gallon capacity)
7. Mixer costs including vaporizer, pump, and air mix unit (standard size units, in million Btu): 3, 7, 10, 14, 20, 30, 40, 50, 60, 70, 80, 90, 100, etc.
8. Installation costs—(including tank piers, electrical, natural gas hookups, fencing, bump post, etc.)
9. Total costs (storage tank, mixer, and installation costs)

Propane Data Processing

Once the data have been collected, the following steps should be taken to price a propane-air backup system.

1. Total installed cost of propane-air system
2. Cost for initial fill of propane
3. Yearly maintenance of system (including semiannual test runs by an outside third-party supplier)
4. Yearly propane loss through vapor loss (at 1/2 of 1% of tank capacity)
5. Total first year cost of propane-air system (installed cost, initial propane fill, yearly maintenance, propane vapor loss, finance charges, etc.)
6. Amount of natural gas used annually—(Mcf/Dth)

7. Current cost of natural gas per unit—(Mcf/Dth)
8. Total current annual cost of natural gas
9. Cost of natural gas per unit if standby propane-air system is installed
10. Total annual cost of natural gas if standby propane-air system is installed
11. Annual savings utilizing standby propane-air system
12. Payback period (in years) for propane-air system

PROPANE-AIR BACKUP SYSTEM CONTRACT

The Supplier's Contract

When the propane-air backup system cost data has been completed; and, if the system appears to be cost effective, the next step is to contract with a supplier. There are generally local suppliers available that are reliable, and sometimes the LDC can provide several names of competent suppliers. Also, a local propane company may install, or know of supplier who installs, these types of systems.

The supplier's contract is very important for both parties. With the use of a fair contract, a large portion of the guesswork or speculation that can accompany a propane-air backup system installation is eliminated.

Following is a listing of items that need to be in any propane-air backup system contract.

1. Design specifications
   a) Propane vaporizer (Btu capacity)
   b) Propane mixer (Flow capacity)
   c) Pressure range (psig)
   d) Air compressor (scfm/psig)
   e) Air compressor motor specifications (hp, phase, volt)
2. Propane storage tank
   a) Capacity
   b) Code compliance
   c) Condition (new or used)
3. Compressor (Size and type)
4. Vaporizer/mixer system (Size and type)
5. Tank support pier specifications
6. Piping for both liquid and vapor lines (size, routing specifications)
7. Site preparation

8. Fire prevention
   a) Fire extinguisher type (Size and quantity)
   b) Hazard requirements
9. Painting (Color, identification coding, etc.)
10. Fence (Size, style, etc. as required)
11. Engineering services, shop drawings, and descriptive data
12. Operating instructions (written, including necessary required drawings)
13. Start-up and adjustment procedures
14. Crane service
15. Testing (all piping must be pressure tested)
16. Prices and terms of sale
17. Contract conditions
18. Taxes
19. Delivery time
20. Freight
21. Warranties

Propane-air Backup System Benefits

Propane-air systems are utilized by many customers to both reduce natural gas costs and the likelihood of interruptions due to circumstances beyond their control.

If natural gas cost and reliability are important, always consider an on-site propane-air backup system addition to the process. The utilization of a propane-air system can have many benefits:

1. Lower natural gas costs since interruptible service can be utilized.
2. Assurance of continuing supply of energy, even if the LDC supply is interrupted and/or curtailed.
3. Potential to negotiate with the LDC for credits and/or payments to curtail natural gas usage during LDC problem periods; i.e., equipment problems, pipeline problems, etc.

SYNOPSIS—THE PROPANE-AIR SYSTEM

Propane-air backup systems may not be at the top of the natural gas cost reduction consideration process, but in the proper environment, a properly designed system can be very beneficial in reducing costs. Since the propane-air system is not like most other systems, it needs to be understood in the context of what it can do to reduce ongoing natural gas costs.

Chapter 16

The Natural Gas Deregulation Process

NATURAL GAS DEREGULATION

The Natural Gas Customer and Deregulation

Deregulation and restructuring, also know as unbundling of services, makes it possible for consumers to purchase service components separately. With complete unbundling, customers can choose their own natural gas supplier; however, the LDC still provides local transportation and distribution services.

The status of deregulation and restructuring of the natural gas industry varies state-by-state as well as by individual customer class such as large commercial, industrial, and residential customers. For the most part, large commercial and industrial customers have had the option of purchasing natural gas separately from transportation and other services for many years.

State regulators and lawmakers responsible for designing and implementing retail, restructuring programs have been slow to implement *choice programs* for residential and smaller-volume commercial customers.

Typically, the concern has been to ensure reliable service. Some state programs do allow residential customers to select natural gas suppliers, while other states have yet to initiate anything. The availability of retail programs, program specifics, participation, and acceptance vary widely across the nation.

The Purchase of Deregulated Natural Gas

Analyzing and understanding the process of selecting a marketer to assist in the purchasing of deregulated natural gas is very important. The entity that generally is utilized to assist a natural gas customer in this process is called a marketer.

Brokers, marketers, and producers are generally defined as *entities authorized by a customer too purchase natural gas in the customer's behalf.*

It is very important for a natural gas customer to realize the legal implications of relationships with natural gas supply marketers. Always keep in mind that when working with a marketer—the marketer can and does legally obligate the natural gas customer (principal) with third parties—natural gas producers and interstate pipelines.

Given these facts, a thorough understanding of the various types of entities—what they provide and what they do not provide—is critical to the establishing of an effective non-LDC natural gas purchase arrangement.

Types of Deregulated Natural Gas

Brokers

A broker provides the function of bringing the natural gas customer (buyer), the natural gas producer, and the interstate pipeline (seller) together. The main and most important distinction of natural gas brokers is that they do not take or assume title to the natural gas they market. They act only as third-party intercessors, and in effect sell for someone else.

Brokers perform the same services as marketers but do not actually take title to the natural as the sell. They are paid a fee for their services by either the customer or seller. This is not to say broker natural gas is unreliable or in anyway different from purchasing from a titled source. The thing to remember is that since title does not pass to the brokers, their warranty as to availability is no better than their source can or will provide.

In general, the fewer steps required to arrive at the actual natural gas production source, the more reliable the supply. Broker-supplied natural gas is not inferior, but remember brokers provide no better title to the natural gas they market than what they have—which is none. If the customer uses a broker, maker sure of the following

1) The broker's source is identified
2) The supply is assured for the duration of your contract
3) The broker's source has title to the natural gas they are selling

In general, it is probably better, both on long-term cost and availability, to contract with either of the other two classes of providers-marketers or producers.

Marketers

Marketers differ from brokers in that *they do take title to the natural gas they sell* to the natural gas customer. Marketers take title to natural gas but do not necessarily have or own producing natural gas wells.

Generally, marketers or marketing affiliates are also known in the natural gas industry as traders. While all of this may seem confusing, it is easy to remember the difference between the marketer category and the broker category is that title to the natural gas passes with the marketer, where it does not pass with the broker.

The difference between marketers and producers is that *producers own the producing natural as wells and marketers ma not.* The marketer category probably is the largest supplier of non-LDC natural gas. As with any group of individual entities, there are good and bad available, so it is important that any contracts negotiated are customer-friendly.

Producers

Producers are as their name implies, *they have title to and also own producing natural gas wells.* They are the original owners of the natural gas and are responsible for getting it out of the ground and into a pipeline.

Some producers market their own product directly to retail customers. Occasionally several producers join together to form an aggregation group who will in turn market the natural gas to retail customers.

Producers are not a dominate force in the retail customer market. In general, they sell their product to either a pipeline, an intrastate natural gas company, or through a broker or marketer who in turn supplies the natural gas retail customer.

Basic Differences between Brokers, Marketers, Producers

Brokers DO NOT TAKE TITLE to nor own producing natural gas wells.

Marketer TAKE TITLE to but do not necessarily own producing natural gas wells.

Producer HAVE TITLE to and own producing natural gas wells.

Generally, there is no reason to limit the choice of a supplier strictly on the basis of the category—broker, marketer, or producer. The criteria for selecting a supplier should be based upon the following quantitative data.

MARKETER TRANSPORTATION CONTRACT— REQUIRED DATA

Reliability of Natural Gas Supply

Have interruptions of natural gas supply occurred? If so, when, for how long and why did they occur?

Does the entity (broker, marketer, producer) have access to multiple natural gas gathering zones so that interruptions can be minimized and natural gas costs remain competitive? If so, where are they located and how many are there?

How Natural Gas Pricing is Determined

Fixed Pricing

The *fixed pricing* method is utilized where a customer requires a fixed non-changing natural gas cost each month. The fixed or firm pricing has nothing to do with deliverability, but refers only to the natural gas cost. Since determining natural gas costs in the future is at best a guess, (generally fixed pricing contracts extend into the future—at least 12 months) a method must be developed to determine the cost that is fair to both the marketer and the customer.

This method must at least base the cost on some benchmark where the future natural gas cost is calculated utilizing an index like the natural gas futures prices as shown on the internet. Forward natural gas futures prices are shown for many months in the future, so there should be no problem when working with a typical 1-year customer contract period.

Generally, the (12) forward month's natural gas futures prices are averaged using individual monthly volume variables to arrive at a weighted average, natural gas cost for each of the (12) forward months. There are other methods to develop these costs but be certain that whatever method is utilized, it at least makes mathematical sense.

Never allow a broker, marketer, or producer to estimate the natural gas price based on their great experience or expertise, because it is likely that either of these techniques will result in higher than necessary natural gas costs for the customer.

Spot Market Pricing

The *spot market pricing* method utilizes a benchmark average pricing index to establish a wellhead natural gas price. There are several reputable accepted indexes from which to choose, three of which are as follows:

- Natural Gas Clearinghouse
- Natural Gas Daily
- Inside FERC

Beware of spot-market index prices originated by a broker, marketer, or producer, because this could result in more costly natural gas to the customer.

INTERSTATE TRANSPORTATION

Routing and Cost

Is transportation or routing available on all interstate natural gas pipelines that intersect or connect to the customer's intrastate LDC city gate receipt point?

Many times more than one interstate natural gas pipeline serves a customer's intrastate LDC. When this is the case, make certain that the broker, marketer, or producer arranges transportation agreements on all applicable interstate pipelines.

By doing this, the likelihood of interstate pipeline interruptions may be reduced since natural gas transportation can be switched between interstate pipelines if required. Also, less costly interstate pipeline transportation rates may be able to be negotiated if competing pipelines are involved.

Are any and all interstate pipeline volumetric discounts passed on to the customer?

Most, if not all, interstate pipelines negotiate natural gas transportation rates based upon volumes of natural gas transported. Always try to select a broker, marketer, or producer that transports a large volume of natural gas to the customer's intrastate LDC.

Customers should require that any interstate pipeline discounts be completely passed through them. The broker, marketer, or producer is compensated by agreed-to contractual charges on a per-unit natural gas transported basis or flat management fee.

Contract Language

Is the contract understandable, and is it customer oriented?

Generally, contracts for any purpose, whether intentionally or unintentionally, tend to favor the contract initiator. It is for this reason that the customer needs to either draft the contract utilized or, at the very least, know the contract, structure, and its specific provisions.

What the customer does not know can be a very expensive proposition.

Does the marketer or producer have permanent title to the natural gas for the entire term of the contract—normally one year?

Permanent title to the natural gas is important to the customer, especially in high usage periods—generally winter months. This is because non-titled natural gas can and will be sold on a first come, highest bid basis. At least if a marketer or producer has title to the natural gas, the customer stands a better chance of receiving their monthly allotment of natural gas.

Permanent title does not guarantee natural gas availability, but it certainly enhances the likelihood of deliverability. Note: marketers and producers are the only two categories that can provide any title to natural gas.

Does the contract provide for the customer to deduct any penalties or similar costs from the monthly billing from the marketer?

Since a broker, marketer, and/or producer act in the customer's behalf, they have the right to contractually obligate the customer to various wellhead, interstate pipeline, and intrastate LDC terms and conditions. In the event they act in the customer's behalf in a manner that results in penalties and/or other similar charges for the customer, some form of redress should be available to the customer.

The contract should provide that the customer is allowed to deduct any penalties or other similar charges from the applicable monthly billing caused by the broker, marketer, or producer. If the contract does not specifically address this issue, it can become a real problem for the customer if, and/or when, a situation of this type happens.

Does the contract specifically identify the point where title to the natural gas passes to the customer?

There are three distinct points in the customer transportation of natural gas process:

1) Natural gas wellhead
2) City gate receipt point of intrastate natural gas LDC
3) Customer burner tip meter point

Any of the three areas can be specified as the point where title passes to the customer. Where title passes becomes important when the customer's burner tip meter point is in a state where a sales tax exemption is available.

If title to natural gas occurs outside of the boundaries of the state, this exemption is generally granted to the customer since, at least contractually, the natural gas purchased was on an interstate (between states) rather than intrastate (within a state) basis.

If this is the case, the state may allow the portion of the transaction that occurs outside of the state to not be subject to state sales tax. This would include all cost from the intrastate LDC's city gate receipt point, back to and including the wellhead natural gas cost.

The only one of the three customer title pass points that can be utilized to not pay state sales tax is the wellhead point. The broker, marketer, or producer should provide guidance to the customer on this matter but ultimately it would be prudent for the customer to consult with either in-house or outside sales tax legal council before deciding where title should pass to them.

Does the contract force majeure *clause (superior or irresistible force) specifically exclude profitability as an exercisable force majeure option by the broker, marketer, or producer?*

Generally, all customer natural gas transportation contracts have a force majeure clause to protect both parties in the event that a part of the contract cannot be performed due to causes which are outside the control of either party and could not be avoided by exercise of due care.

However, there have been occasions where this clause has been exercised simply due to the profitability or lack thereof of completing of the contractual obligation.

Although this action could probably not be defended legally, it can put the customer in a position of having no natural gas or at least no reasonably priced natural gas. It is best to have a force majeure contract clause that has specific wordage precluding profitability as a valid exercisable option.

Customer Service

Is incremental data provided on monthly billings?

Good incremental base data are critical since the customer's responsibility of the natural gas transportation is much greater in this transaction than if they were purchasing natural gas directly from their intrastate LDC. It is generally the customer who has to determine what data they require on a monthly basis.

The items that need to be individually recorded on the monthly billing should, at a minimum, include the following items:

1) Quantity of natural gas used by the customer at their burner tip meter point
2) Quantity of natural gas shipped from the wellhead meter point
3) Cost of natural gas per unit shipped from the wellhead meter point
4) Broker, marketer, or producer charge per unit or management fee for natural gas shipped from the wellhead meter point
5) Actual interstate natural gas transportation fee that was able to be negotiated on customer's behalf
6) Miscellaneous interstate pipeline fees that may be present—Take or Pay, Gas Research Institute, miscellaneous
7) Interstate pipeline shrinkage factor
8) Listing of any/all taxes—federal, state, etc.
9) Billing recap that details customer's total savings for the specific billing period, as well as total savings year-to-date or other period as requested by the customer

The Purchase of Deregulated Natural Gas

Although there may seem to be a large amount of information to be evaluated and included in the contract, all of it is required if the customer is to be assured they will receive the most reliable, lowest cost natural gas.

It is much better to spend the time required initially to structure the contract and the monthly billing data requirements correctly than to experience a problem later. Trying to correct a problem is never as satisfactory as is doing the job correctly in the first place.

To better understand the incremental steps in the transportation of natural gas for the customer, the following nine items are listed. All of these items, if applicable, should be individually itemized on the broker, marketer, or producer billing to the customer.

Necessary Items on Customer Billing

Pricing Point #1—Wellhead

1) Wellhead cost of natural gas
2) Third-party charge
3) Interstate transportation cost
4) Natural gas shrinkage cost
5) Take or pay surcharge
6) Miscellaneous charges

Pricing Point #2—Interstate LDC City Gate
(Total of Items 1-6 plus Items 7-8)

7) Intrastate transportation cost
8) Natural gas shrinkage cost
9) Miscellaneous charges

It is very important that the customer understand all of the cost components in the transaction in order to evaluate the cost effectiveness of each item that affects their burner tip cost. Also, a customer should always require incremental documentation of all costs in the natural gas transaction on an individual billing basis.

This should not present a problem to the broker, marketer, or producer since all of the required cost data is available for them to be able to calculate the costs for their portion of the transaction—wellhead to intrastate LDC city gate point. If a broker, marketer, or producer is not willing to provide incremental data, it would probably be better to work with one who will provide the required data.

- It is the customer's responsibility to know and control their natural gas costs.

- The broker, marketer, or producer works for the customer, not the natural gas owner, interstate pipeline, or the customer's LDC.

- Any problems are the customer's problems to pay for and correct.

SYNOPSIS—THE NATURAL GAS DEREGULATION PROCESS

A thorough understanding of the natural gas deregulation process is basic to an effective energy cost reduction process. With the opportunities that are available in the area of deregulated natural gas, many commercial and industrial natural gas customers utilize the process. The use of the deregulation process reduces cost but does require the customer to have knowledge of how the deregulated provider process works.

Chapter 17

The Natural Gas Contract Process

CONTRACT PROCESS

The Efficient Cost-effective Natural Gas Purchase Arrangement

This chapter explains the various contract processes that the natural gas customer is involved in when utilizing a provider (broker, marketer, and/or producer) to obtain natural gas in their behalf. The items addressed contain the keys to an efficient cost-effective, customer-initiated natural gas purchase arrangement. Since a written contractual contract between customer and their provider is required, it is critical that it be structured in a manner that is fair for both parties.

Generally, a provider will present the customer with a written contract that in many cases is more beneficial to the provider than the customer. This is not to say a provider-originated contract is in any way illegal, but it may favor or benefit the provider more than the customer.

It is for this reason that the contract process will be detailed in this chapter. The contract detailed here is generic or general in nature and will probably never be utilized in its entirety for any actual transaction. However, there are very important parts of this contract that should always be incorporated in any customer/provider contract. A thorough understanding of the contract process is required by the customer to maximize natural gas savings that are available.

When negotiating any written contract, always utilize appropriate legal council to assure that any company, state or federal guidelines, and/or laws are not being violated. The contract that will be explained in this chapter is fair to both the customer as well as the provider.

Its provisions should cause no problems for either party, in that it is simple and straightforward by design in its format. However, since it is impossible to anticipate every situation that could arise between a customer and a provider, the contract shown in this chapter should only be used as a guide or outline for the actual establishment of a customer/provider contract.

Figure 17-1. Natural Gas Contract Process

NATURAL GAS SALE CONTRACT

Between

____

____

And

____

____

Contract No.__________________

Page 1

Figure 17-1. (*Continued*)

INDEX

Page 2

Figure 17-1. (*Continued*)

NATURAL GAS SALE CONTRACT

THIS CONTRACT, made and entered into this __*(day)*______*(month)*, _____*(year)*.

CUSTOMER

BY: ____

With Principal Offices at: ____________________________

____

BETWEEN

PROVIDER ***(Broker, Marketer, and/or Producer)***

BY: ____

With Principal Offices at: ____________________________

____

1. Definitions

1.1 Dth means (1) one million British Thermal Units.

1.2 EXPIRATION DATE refers to 8:00 a.m. on the date that is on the *(1)* year after PROVIDER first delivers natural gas to CUSTOMER hereunder.

1.3 PIPELINE QUALITY NATURAL GAS is natural gas that meets the quality and Btu requirements of the receiving pipeline at Point of Delivery.

1.4 LOCAL DISTRIBUTION COMPANY *(LDC)* refers to CUSTOMER'S LDC.

1.5 POINT OF USAGE is the location at which CUSTOMER will consume or otherwise use said natural gas.

Page 3

Figure 17-1. (*Continued*)

2. Purchase and Sale

2.1 PROVIDER agrees to sell to CUSTOMER and CUSTOMER agrees to purchase from PROVIDER up to *(Maximum Monthly Volume)* Dth of natural gas *(the Maximum Monthly Volume)* hereinafter each month during the term of this Contract. Said natural gas will be ordered by PROVIDER in CUSTOMER'S behalf as provided for in this Section. Delivery and acceptance shall be at Point of Delivery *(as defined hereinafter)*.

2.2 CUSTOMER may use any volume up to the *Maximum Monthly Volume* for any specific month or months.

2.3 CUSTOMER and PROVIDER may mutually agree, at any time, up to the delivery of additional quantities of natural gas *(over and above the Maximum Monthly Volume)* if PROVIDER has such additional quantities available.

2.4 If in any month PROVIDER fails to tender an amount of natural gas equal to the Maximum Monthly Volume, or if in any month PROVIDER fails to tender the volume nominated up to the Maximum Monthly Volume for that month, CUSTOMER may at its sole option elect to terminate this Contract.

2.5 PROVIDER warrants that it has permanent title to natural gas for the entire term of this Contract hereunder and the right to sell same.

3. Term of Contract

This Contract shall become effective upon execution and shall remain in effect for an initial term of twelve *(12)* months after natural gas is first delivered hereunder.

4. Price of Natural Gas

FIXED PRICE METHOD

4-A. Price of Natural Gas – Fixed Price for (_____) Month Period

4-A.1 The purchase price shall be $______per Dth delivered at Point of Delivery during the initial (_______) month period.

4-A.2 The PROVIDER shall provide to the CUSTOMER a new purchase price at least forty-five *(45)* days prior to expiration of the initial (_________) month period *(or of any extended period)*. Any such new price shall be evidenced by a written amendment to this Contract executed by both parties. Failure to agree upon a price per Dth shall permit either party to cancel this Contract effective at the end of the then current (______) month period, by providing thirty *(30)* days prior written notice of such cancellation to the other.

Page 4

Figure 17-1. (*Continued*)

Fixed Price Method Continued

4-A.3 CUSTOMER will *(in addition to paying the prices established in 4-A.1)* pay all state and local sales, use, and public utility taxes *(associated with the sale contemplated by this Contract)*; and, all other costs associated with the transportation, handling ownership, sale distribution and use of natural gas after acceptance by CUSTOMER at Point of Title transfer as set forth herein.

4-A.4 CUSTOMER shall have no responsibility to pay production or severance taxes or to make royalty or other payments due out of production.

4-A.5 The prices established in *4-A.1*, shall apply to all natural gas delivered pursuant to this Contract.

SPOT MARKET METHOD

4-B. Price of Natural Gas – Spot Market

4-B.1 The purchase price shall be determined by the Spot Market Pricing Guide chosen, per Dth in effect on date of nomination.

4-B.2 Not used with Spot Market Pricing.

4-B.3 CUSTOMER will *(in addition to paying the prices established 4-B.1)* pay all state and local sales, use, and public utility taxes *(association with the sale contemplated by this Contract)*; and, all other costs associated with the transportation, handling ownership, sale, distribution, and use of natural gas after acceptance by CUSTOMER at Point of Title transfer as set forth herein.

4-B.4 CUSTOMER shall have no responsibility to pay production or severance taxes or to make royalty or other payments due out of production.

4-B.5 The prices established in *4-B.1* shall apply to all natural gas delivered pursuant to this Contract.

5. Deliveries

Deliveries shall commence by no later than fifteen *(15)* days after execution of this Contract and of all transportation and delivery Contracts required for transporting said natural gas to Point of Usage. If deliveries do not so commence, CUSTOMER may terminate this Contract.

Page 5

How to Establish a Fixed Price for Natural Gas

(This is not part of the contract)

Generally, a fixed price method is more costly than the spot market method because it can be difficult to determine the cost of natural gas at some point in the future.

Typically, fixed price contracts are for 3, 6, 9, 12 months or longer. The longer the term, the more speculative is the cost of natural gas. Unless, for some reason, a known natural gas cost is required on a month-by-month basis for some future period of time, it is probably less costly to utilize the spot market or monthly pricing scenario for natural gas purchases.

If a fixed natural gas price is required, then utilize some benchmark to at least make the process logical and not pure speculation. One benchmark that could be utilized would be the NYMEX index of natural gas futures prices. This data is available every weekday in various publications or on the internet.

Establishing a Spot Market Price for Natural Gas

This method utilizes a reliable reference price per MMBtu on a monthly basis to determine the unit natural gas cost for that month. There are at least four reliable reference price indexes available:

1. NYMEX Index
2. Inside FERC
3. Natural Gas Daily
4. Natural Gas Clearinghouse

Any of these indexes are acceptable as they fairly represent actual natural gas costs on a monthly basis. The spot market method is simple to utilize since the applicable monthly index cost chosen times the monthly volume, equals the total natural gas cost for that month.

Figure 17-1. (*Continued*)

6. Billing and Payments

6.1 Billing shall be rendered monthly. The monthly billing period *(Billing Period hereinafter)* shall end on the first day of each calendar month. The first billing period will end on the last day of the month in which deliveries commence, whether or not a full month of deliveries is involved. CUSTOMER will be required to accept and to pay each month for the Ordered Monthly Volume, pursuant to Section 2; and, for all additional quantities of natural gas ordered and delivered by mutual contract of the parties pursuant to Section 2.3 of this Contract.

6.2 Monthly billings will be made based on the monthly volumes ordered by CUSTOMER pursuant to Section 2, and corrections to billings *(if any)* will be reflected in the next billing rendered after the need for correction is discovered. Payment *(or credit)* for corrected monthly billings will be made and will be reflected in the next due payment. Corrections will not be made more than one *(1)* year after the original billing date.

* 6.3 CUSTOMER shall pay all amounts due less any penalties incurred by PROVIDER'S incorrect nominations, and/or balancing procedures which result in Local Distribution Company penalties, surcharges, and/or supplying of natural gas at a cost in excess of natural gas that could have been purchased under provisions of this contract.

6.4 Payments shall be made by electronic transfer, within five *(5)* days after receipt of a billing from PROVIDER; or, by the fifteenth *(15th)* day following the month of delivery, whichever shall last occur. PROVIDER'S billing may be submitted by e-mail, fax, telegram, or other written instrument.

Payments will be made into an Escrow Account at *(any bank selected by* PROVIDER*)*. CUSTOMER agrees to execute any necessary instructions to the Escrow Bank *(as directed by* PROVIDER*)*, so long as said instructions do not change or enlarge interest equal to prime rate at the Escrow Bank, plus an amount agreed-to by the parties shall be paid on all late payments. CUSTOMER shall, in addition, be responsible for paying for collection costs and reasonable attorney fees incurred by PROVIDER in its efforts to collect delinquent payments.

6.5 If failure to pay shall continue for ten *(10)* days after receipt of a billing by CUSTOMER, PROVIDER may, in addition to any other remedies available, suspend further deliveries to CUSTOMER until all amounts due are paid.

6.6 It is contemplated that credit circumstances and requirements of each CUSTOMER will be determined after execution of this Contract. This section *(6.6)* is reserved for any mutually satisfactory provisions that may result from said determination. In the event no mutually satisfactory credit arrangements are made, PROVIDER may terminate this Contract without further obligation hereunder

* *Mandatory provision in any natural gas contract*

Page 6

Figure 17-1. (*Continued*)

7. Transportation Contracts

Point of Usage is identified in *Exhibit A* hereto. PROVIDER, as requested by CUSTOMER, will negotiate and arrange all transportation and delivery contracts with pipeline companies, and with Local Distribution Companies *(LDCs)*. CUSTOMER recognizes it may be called upon to aid and assist PROVIDER in the negotiation of said Pipeline and LDC Contracts. CUSTOMER recognizes that it will be a required signatory party to such transportation and/or delivery contracts, and agrees to execute same.

8. Title to Natural Gas

8.1 Title to natural gas sold pursuant to this Contract will pass to CUSTOMER at Point of Delivery. Point of Delivery is identified in *Exhibit B* hereto, which is incorporated herein by reference.

8.2 PROVIDER may change Point of Delivery; and, will negotiate and arrange all necessary new transportation and delivery contracts.

8.3 In the event of such change in Delivery Point, the then current price of natural gas *(established by Section 4)* will be changed to a price which when added to the new total of transportation and delivery costs *(from said new Point of Delivery to the location identified in Exhibit A hereto)* exactly equals the price plus all transportation and delivery costs *(from the old Point of Delivery)* of natural gas delivered during the prior monthly Billing Period.

8.4 Price adjustments *(if any)* during months subsequent to the establishment of new Point(s) of Delivery will be made as provided in *Section 4*.

9. Force Majeure

9.1 All obligations of the parties to this Contract *(except for he payment of money for natural gas delivered)* shall be suspended while and for so long as compliances prevented in whole or in part by an act of God, stroke, lockout, war, civil disturbance, explosion, breakage, accident to machinery or pipeline, failure of well or sources of natural gas supply, federal or state or local law, inability or refusal of any pipeline or Local Distribution Company *(LDC)* to accept natural gas for delivery, or otherwise transport the default of any party to Other Contracts *(other than CUSTOMER or* PROVIDER*)*, or by any other cause beyond the reasonable control of CUSTOMER or PROVIDER.

9,2 No part of this Force Majeure clause shall be construed to provide for the discontinuance of natural gas delivery by the PROVIDER due to natural gas not being available at a favorable cost to the PROVIDER. The PROVIDER shall be obligated to provide natural gas to the CUSTOMER, subject to other provisions of the Contract, at the agreed-to price, without regard to the profitability to the PROVIDER. Ability of the PROVIDER to realize a profit on the natural gas transaction shall not be considered as a Force Majeure condition providing for suspension of the obligation of the PROVIDER to provide natural gas to the CUSTOMER.

Figure 17-1. (*Continued*)

10. Notices

Any notice or other communication required or desired to be given to any party under this Contract shall be in writing and shall be deemed given when: (a) delivered personally to that party; or, (b) delivered by the United States mail, certified postage prepaid, return receipt requested, or delivered by overnight delivery, or delivered by overnight delivery, e-mail, tax, or courier return receipt requested, addressed to that party at the address specified for the party earlier in this Contract or at any other address hereafter designated by that party by written notice.

PROVIDER	CUSTOMER
PAYMENTS:	***PAYMENTS:***
Name:__________	**Name:**__________
Address:__________	**Address:**__________
Attention:__________	**Attention:**__________
Correspondences:__________	**Correspondences:**__________

11. Miscellaneous

11.1 If any provision of this Contract is found to be invalid or unenforceable *(other than the requirement concerning payment for natural gas delivered)*, it is intended that the balance of this Contract remain in full force and effect.

11.2 This Contract is intended for the exclusive benefit of the parties to this Contract and their respective heirs, successors, and assigns; and, may not be assigned without written approval of the non-assigning party. Northing contained in this Contract shall be construed as creating any rights or benefits in or for any third party.

11.3 The obligation of CUSTOMER(s), if more than one *(1)*, shall be joint and several.

11.4 This document *(including Exhibits and Addenda, if any)* contains the entire Contract between the parties and supersedes all entire Contract between the parties and superseded all prior or contemporaneous discussions, negotiations, representations, or contracts relating to the subject matter of this Contract. No change to this Contract shall be made or shall be binding on any party unless made in writing and signed by each party of this Contract.

11.5 This Contract shall be binding upon and shall inure to the benefit of the parties hereto, their respective heirs, successors, and assigns.

11.6 This contract is expressly made subject to all present or future valid rules, orders or regulations of duly constituted governmental authorities having jurisdiction over the subject matter hereof.

11.7 The failure, by either party hereto, to act in the event of default shall not constitute a waiver of the right to so act unless otherwise provided herein.

11.8 <u>QUALITY</u>: All natural gas delivered hereunder shall meet the quality and heat content requirements of the applicable transporting pipeline.

Page 8

Figure 17-1. (*Continued*)

11.9 PROVIDER By execution of this contract, CUSTOMER hereby appoints PROVIDER to act as its exclusive PROVIDER to - (a) manage, nominate, and schedule transportation service with the LDC from the Delivery Point(s) to the Business Meter(s) in accordance with CUSTOMER'S LDC service contract or rate schedules and this Contract; (b) communicate directly with the LDC to balance scheduled and actual receipt and delivery of CUSTOMER'S natural gas pursuant to CUSTOMER'S LDC service contract and this Contract; and, (c) do all other acts necessary to perform the duties set out above. The agency herein created is a limited agency, and PROVIDER'S duties are specifically limited to those set forth above and no other duties of any kind or natural gas are imposed on PROVIDER including without limitation any duties which may otherwise arise by operation of law. CUSTOMER hereby acknowledges and agrees that during the term of this Contract, PROVIDER is free to perform any and all agency services pursuant to this contract with any third party. CUSTOMER shall defend indemnify and hold PROVIDER harmless from and against any and all costs, claims, damages, expenses, fees, penalties, suits, and/or actions of any kind, character, or nature including reasonable attorneys' fees and court costs, that may arise out of or be connected with PROVIDER'S acts as PROVIDER for CUSTOMER except for those acts arising from PROVIDER'S gross negligence.

11.10 CUSTOMER CREDIT: CUSTOMER shall promptly notify PROVIDER of any material changes to its financial condition during the Term. CUSTOMER shall provide credit information sufficient for PROVIDER to make reasonable inquiry into CUSTOMER'S creditworthiness and ability to continue performing its obligations under this contract within five *(5)* business days after PROVIDER'S request. If CUSTOMER fails to so provide such requested information, or in PROVIDER'S good faith determination, there has been a material adverse change in the credit status or financial condition of CUSTOMER, the CUSTOMER and PROVIDER shall mutually agree on a specific form of credit assurance *(i.e., letter of credit, cash, or a guarantee from a creditworthy guarantor – as determined by* PROVIDER*)*. CUSTOMER shall provide such credit assurance to PROVIDER within five *(5)* calendar days after PROVIDER'S request.

11.11 TAXES AND ROYALTIES: PROVIDER shall pay or cause to be paid all royalties and other sums relating to the production and transportation of natural gas to the Delivery Point(s); provided, however, that any increase in applicable taxes or transportation fees relating to the production and transportation of natural gas to the Delivery Point(s), and any increase in PROVIDER'S costs due to changes in applicable federal or state laws or regulations occurring after the Effective Date will be paid by CUSTOMER and reflected in the invoice as either an adjustment in price or as an additional item, at PROVIDER'S sole discretion. CUSTOMER shall be responsible for any costs, fees, and taxes of any kind which are levied upon or attributable to the natural gas purchased and sold hereunder after title has passed from PROVIDER to CUSTOMER, including but not limited to the utility shrinkage, franchise fees, lost and unaccounted for charges, utility users' tax, intrastate transportation charges and surcharges of every kind and nature.

Page 9

Figure 17-1. (*Continued*)

Any increase in applicable taxes will be paid by CUSTOMER and reflect in the invoice as either an adjustment in price or as an additional item, at PROVIDER'S sole discretion. CUSTOMER shall be responsible for sales, usage, and any other related taxes. CUSTOMER shall provide PROVIDER with any applicable certificate or other documentation of sales or use tax exemption; and CUSTOMER shall be liable for any sales or use tax and associated interest or penalties assessed against PROVIDER due to CUSTOMER'S failure to timely provide or properly complete any such certificate or documentation.

11.12 MEASUREMENT: The measurement, calculation, and computation of the volume and quantity of natural gas delivered by PROVIDER at the Delivery Point(s) shall be determined in accordance with the specifications set forth in the transportation rate schedule of the interstate pipeline delivering natural gas for the account of PROVIDER or in accordance with the terms of the transportation contract between PROVIDER and the transporter delivering natural gas for the account of PROVIDER if other than an interstate pipeline.

11.13 TRANSPORTATION: PROVIDER shall pay for the transportation of the natural gas purchased and sold hereunder to the Delivery Point(s), and CUSTOMER shall pay for transportation of the natural gas purchased and sold hereunder from the Delivery Point(s).

11.14 TITLE; WARRANTIES; INDEMNITY: Title to the natural gas bought and sold hereunder and risk of loss shall transfer from PROVIDER to CUSTOMER as provided in this Section. PROVIDER shall have title to, responsibility for, and assume any liability with respect to the natural gas prior to its delivery to CUSTOMER at the specified Delivery Point(s). CUSTOMER shall have title to, responsibility for, and assume any liability with respect to said natural gas at the Delivery Point(s) and after its delivery to CUSTOMER at the Delivery Point(s). PROVIDER agrees to indemnify CUSTOMER and save it harmless from all losses, liabilities, or claims including attorneys' fees and costs of court *(Claim)*, from any and all persons, arising from or out of claims of title, personal injury or property damage from said natural gas or other charges thereon which attach before title passes to CUSTOMER. CUSTOMER agrees to indemnify PROVIDER and save it harmless from all Claims, from any and all persons, arising from or out of claims regarding payment, personal injury or property damage from said natural gas or other charges thereon which attach after title passes to CUSTOMER. PROVIDER warrants that it has the right to sell the natural gas hereunder and that such natural gas is free from liens and material adverse claims of any kind.

11.15 DISCLAIMER OF WARRANTIES: Except for the express representations and warranties set forth herein, PROVIDER expressly negates any representation or warranty, written or oral, express or implied, including any representation or warranty with respect to conformity to models or samples, merchantability, or fitness for any particular purpose. CUSTOMER acknowledges that it has entered into this Contract based solely upon the express representations and warranties contained herein and, subject to such express representations and warranties, CUSTOMER otherwise accepts the natural gas *as-is* and *with all faults*.

Page 10

Figure 17-1. (*Continued*)

11.16 <u>THIRD-PARTY BENEFICIARIES; RELATIONSHIP OF THE PARTIES:</u> There are no third-party beneficiaries to this contract including, but not limited to, the LDC, and none are intended by the parties. Any financial transaction(s) entered into by CUSTOMER related to this contract are deemed made at CUSTOMER'S election in the exercise of its independent judgment to assume any risk associated with the same, and PROVIDER disclaims any intent to advise or influence CUSTOMER.

11.17 <u>LIMITATION OF LIABILITY</u>: Neither party shall be liable to the other party for any special, indirect, incidental, consequential, or punitive damages, including but not limited to lost profits, of any character, irrespective of whether claims or actions for such damages are based upon contract, warranty, negligence, strict liability, or any other remedy at law or equity.

11.18 <u>WAIVER</u>: Either party's failure to insist upon strict performance of any provision herein shall not constitute a waiver of the right to require such performance in the future. In the event that any provision(s) contained herein should be found to be unenforceable in any respect, such finding shall not affect any other provision of this contract, and this contract shall then be construed as if such unenforceable provision(s) had never been contained herein.

11.19 <u>AMENDMENT</u>: PROVIDER reserves the right to modify this contract upon written notice to CUSTOMER in the event of substantive changes in CUSTOMER'S LDC transportation rate schedule, LDC operational provisions and/or tax law. CUSTOMER and PROVIDER agree that this sale is subject to any and all existing and future valid laws, orders, directives, rules, and regulation of the regulatory bodies having jurisdiction over the parties and this transaction.

11.20 <u>TERMINATION</u>: In addition to all of its rights and remedies hereunder and at law and in equity, PROVIDER may terminate all obligations owed CUSTOMER as its PROVIDER, suspend deliveries hereunder, and/or terminate this contract, without liability, at any time, after the occurrence of any of the following conditions, by written notice to CUSTOMER given before CUSTOMER cures the same – (a) CUSTOMER fails to make any payment due hereunder within thirty *(30)* calendar days after it is due; (b) CUSTOMER misrepresents its financial condition when providing credit information or financial assurances as requested under Section 3; (c) CUSTOMER breaches any of the other material terms and conditions of this contract and fails to cure such default within thirty *(30)* calendar days from receipt of written notice from PROVIDER; or, (d) CUSTOMER makes a general assignment for benefit of its creditors, is insolvent, has a receiver appointed because of insolvency, files bankruptcy, or has a petition for involuntary bankruptcy filed against it. In any such event, PROVIDER may terminate this contract taking reasonable steps to notify CUSTOMER'S LDC and CUSTOMER of the need to arrange for alternate natural gas supply. CUSTOMER has a right to terminate this contract prior to the end of the term, if PROVIDER breaches any of the material terms and conditions of this contract and fails to cure such default within thirty *(30)* calendar days from receipt of written notice from CUSTOMER. CUSTOMER may thereafter terminate this contract by written notice given to PROVIDER.

Page 11

Figure 17-1. (*Continued*)

11.21 GOVERNING LAW: This contract shall be constructed according to the laws of the State of ______________ without giving effect to principles of conflicts of laws.

11.22 CONFIDENTIALITY: Each party shall keep the terms and conditions of the contract confidential except as may be required in order to effectuate the transportation and delivery of natural gas to be sold hereunder or to meet the lawful requirement of any regulatory body or court having jurisdiction. PROVIDER and CUSTOMER agree to consult with one another prior to issuing any press release or similar public statement relating to this contract or the services contemplated hereunder. CUSTOMER does not authorize PROVIDER to use its name in CUSTOMER lists and other promotional materials that PROVIDER may develop from time to time except upon specific prior written CUSTOMER authorization.

11.23 SUCCESSORS AND ASSIGNS; ASSIGNMENT: This contract shall be binding upon the permitted successors and assigns of the parties. Neither party may directly or indirectly assign, transfer, or otherwise convey any of its rights or obligations under this contract.

Page 12

Figure 17-1. (*Continued*)

SIGNATORIES TO THE CONTRACT

CONTRACT NO.______________

EXECUTED: This _____*(day)* **of** _________*(month)*, _____*(year)*

PROVIDER *(Broker, Marketer, and/or Producer)*

By:_________________________________

Printed:_____________________________

Title:_______________________________

Date:_______________________________

CUSTOMER

By:_________________________________

Printed:_____________________________

Title:_______________________________

Date:_______________________________

Page 13

Figure 17-1. (*Continued*)

ACKNOWLEDGEMENTS

PROVIDER

State of:____________________ County of:____________________

Executed and acknowledged by____________________, as the act and deed of ____________________. Before the undersigned Notary Public this________*(day)* of ____________*(month,* __________*(year)*.

____

Notary Public in and for

__________ County, __________State

My Commission Expires:

____

CUSTOMER

State of:____________________ County of:____________________

Executed and acknowledged by____________________, as the act and deed of ____________________. Before the undersigned Notary Public this ________*(day)* of ____________*(month,* __________*(year)*.

____

Notary Public in and for

__________ County, __________State

My Commission Expires:

____

Page 14

Figure 17-1. (*Continued*)

EXHIBIT 'A'

POINT OF USAGE

COMPANY NAME:____

FACILITY ADDRESS TO RECEIVE NATURAL GAS:

____

____

____

LOCAL DISTRIBUTION COMPANY (LDC):________________________

EXHIBIT 'B'

POINT OF DELIVERY *(To be completed by* PROVIDER*)*

POINT OF DELIVERY: *(All interconnect points on–)*

*(Name)*____________________________ Pipeline onshore and offshore

RECEIVING NATURAL GAS PIPELINE:

*(Name)*____Pipeline

EXHIBIT 'C'

PROVIDER'S CHARGE / MANAGEMENT FEE

PROVIDER'S Charge / Management Fee enumerated in this contract shall bo ao follows:

Charge per unit or time period:

$____

Page 15

EXPLANATION OF FIGURE 17.1—
THE NATURAL GAS CONTRACT PROCESS

Natural Gas Sales Contract

The natural gas sales contract outlines the parties to the contract, the effective dates, and the addresses of both the PROVIDER and the CUSTOMER (buyer).

1. Definitions

1.1 Dth—(1,000,000 Btu)

This item identifies the measurement criteria used for delivery of the natural gas. The Dth is the standard term used. On occasion, Mcf (1,000 cubic feet) or therm (100,000 Btu) are used. Any of these units of measurement can be used as long as the parties to the contract identify, understand, and agree to them.

1.2 Expiration date

This determines the date upon which this contract becomes null and void.

1.3 Pipeline quality natural gas

Since natural gas quality varies greatly, it is important that the natural gas be of a quality acceptable by the receiving pipeline as specified in the contract as indicated here.

1.4 Local distribution company (LDC)

This refers to the customer's intrastate natural gas LDC that will be delivering the contracted natural gas.

1.5 Point of usage

This identifies where the natural gas will be consumed.

2. Purchase and sale

This section constitutes the contract between the provider and customer in its entirety. It outlines the conditions under which the natural gas is accepted, and limits the customer's responsibility to the provider. Note: the provider is responsible for nomination of natural gas in customer's behalf.

2.1 **Sets forth the maximum quantities of natural gas the provider is obligated to provide.**
This does not obligate the customer to any specific volume purchases.

*2.2 **Sets forth the customer's options relating to natural gas volumes ordered**
Basically, the customer has the right to use up to the maximum quantity of natural gas set forth in 2.1.

2.3 **Additional quantities of natural gas**
This item allows the customer to purchase quantities of natural gas in excess of the maximum amount stated in 2.1 if the provider can obtain it.

*2.4 **Provider's failure to deliver natural gas**
This allows the customer to terminate the contract if the provider cannot or does not deliver the quantity of natural gas required.

*2.5 **Title to natural gas**
Title to natural gas is important since actually taking title is the only way to assure that the specified quantities will be available when needed.
***NOTE**: The words *permanent* and for the *entire term* of this contract are important since the term title to natural gas may not apply to the entire length of the contract. Permanent title does not guarantee deliverability of natural gas, but it does assure that interruption will not occur for other than actual physical supply or transportation constraints.

3. **Terms of contract**
This section defines when the contract begins and ends.

4. **Price of natural gas**
Two pricing services are provided—Fixed Price and Spot Market Price.

4-A Fixed Price (Option #1)
This annual pricing method is many times utilized by natural gas

customers. Its only advantage is that the natural gas cost is known for a 12-month forward period. Usually, this pricing method results in natural gas costs that are more expensive than Option #2—Spot Market Price.

The fixed price method is generally more costly than the spot market method because the future cost of natural gas has to be estimated. Typically, fixed price contracts are for (12) months. The longer the term, the more speculative is the cost figure. Unless a fixed, monthly natural gas cost is required, it is probably less costly to utilize the spot market option.

If a fixed natural gas price is desired, then utilize some benchmark to make the process as logical as possible. One reliable benchmark is the NYMEX strip of natural gas futures prices. This information is available every business day on the internet.

Although this fixed price method probably may result in natural gas that is more expensive for the customer, it may apply to customers who must lock in their costs for some future period of time.

It is important to remember that fixed costs do not guarantee a supply of natural gas. Fixed pricing means only that if the natural gas is available, it will be sold at the agreed-upon price.

The use of a pricing guide allows providers to hedge (purchase or sell natural gas futures) to ensure they will not pay more for the natural gas than they can get selling it to the customer. However, fixed natural gas future pricing may not be the way to maximize the customer's savings potential. The spot market price may result in uneven natural gas costs on a monthly basis; but on an annualized basis, the natural gas costs may be lower.

4-B Spot Market Price (Option #2)

In this method, a spot market price index is selected by generally using Inside FERC, Natural Gas Clearinghouse, or Natural Gas Daily to fix the wellhead natural gas price on a monthly basis. This spot market method generally results in lower, overall natural gas costs than the fixed price method.

Basic Price Data for 4.A/B

4.A/B.1—Natural gas price per specified quantity (Dth, Mcf, therm, etc.)

This item sets the agreed-upon price for the natural gas during the term of the contract (either monthly or annually).

4.A/B.2—New purchase price notification by provider to customer

This item details how a new purchase price will be negotiated at the end of the agreed-to period.

Note: Either the provider or customer can cancel the contract if an agreement cannot be reached on a new price.

Do not sign a contract that contains a clause, such as "Provider shall have the right to match the price and term of any contract offer(s) to the customer for such subsequent periods."

This says that the current provider has the right to match any natural gas price. If a provider cannot arrive at a competitive price, then in all probability, the customer needs to get another provider.

4.A/B.3—Customer-assumed taxes and other costs

Since the provider is offering only natural gas, it is not their responsibility to pay any sales, local, LDC, or other taxes. And it is not the provider's responsibility to pay for any transportation, handling, or distribution charges after transferring title to the customer. (Title transfer is detailed in 8.1)

4.A/B.4—Provider-assumed taxes and other costs

The provider assumes responsibility for all taxes and other costs so long as they have title to the natural gas. Simply put, whoever has title to the natural gas assumes the related costs—taxes or other expenses.

4.A/B.5—Applicability of natural price established

The price established applies to all natural gas delivered by the provider, whether under or over the specified maximum monthly volume. This clause means that the provider cannot increase the natural gas charges because of monthly order volume variations.

5. Deliveries

This section outlines the provider's responsibilities for beginning initial deliveries and the customer's remedies if delivery does not begin as outlined.

6. Billing and payment

6.1 Billing periods

This item sets forth the billings periods—normally monthly.

6.2 Monthly billing amount computation

This item details what constitutes a monthly billing amount.

***6.3 Payments of amounts due by customer**

This item outlines when the customer must satisfy monthly billings posted by the provider. Also identified are those items that the customer may deduct from the payment due, caused by the provider's incorrect nomination or balancing procedures.

6.4 Delinquent-payment, late charges

The interest penalty amount and the provider's legal recourse in the case of delinquent payments by the customer are outlined here.

6.5 Suspension of delivery for delinquent payments

Detailed here is the provider's recourse in the event of nonpayment in excess of a specified period of time by the customer.

6.6 Determination of credit worthiness of customer

This item covers any special provisions that might be prompted by an unusual credit circumstance of a customer. There must be a mutual agreement by both parties before any provisions can be attached to this item.

7. Transportation contracts

This section covers those situations in which the customer wants the natural gas delivered. It outlines the customer's responsibility to assist the provider in obtaining transportation.

8. Title to natural gas

***8.1 When title passes to customer**

This item defines when title passes to the customer and when the customer becomes responsible to taxes and other costs (when title passes from provider to customer).

Title can pass at three distinct locations: 1) wellhead meter point, 2) intrastate natural gas LDC city gate meter point, and 3) customer burner tip meter.

From the customer's viewpoint, the best place for title to pass is at the wellhead meter point. This is because state sales tax will not normally be charged on the portion of the transaction from wellhead to intrastate LDC city gate meter point if title passes to the customer at the wellhead location. For customers in states that levy sales taxes on natural gas purchases, title pass-point is an important consideration.

8.2 Delivery point changes

This allows the provider to change delivery points as needed. Changes could be mandated by availability of natural gas, mechanical pumping, handling problems, or severe weather conditions.

8.3 Delivery point change effect on natural gas price to customer

This item simply says that changes in delivery points will not affect the customer's delivered cost.

8.4 Price adjustments

If a delivery point change results in higher costs to the provider, the effect on the customer depends upon whether a spot market method or a fixed price method is utilized.

A spot market method means that any cost differentials will be passed through to the customer in the month that the change is made. In the case of a fixed price method, any cost differentials will be passed through to the customer only after the fixed price period has expired. Naturally, if catastrophic or unforeseen circumstances drastically affect the provider's ability to meet the agreed-to price, it would be in the best interest of the customer to consider adjusting the purchase price with the provider. Reasonableness is the test that must prevail.

9. Force Majeure *(Superior or Irresistible Force)*

9.1 Responsibility limits

This limits the responsibility of both parties in circumstances over which neither have control—acts of God, war, accident, etc. If the provider cannot deliver or the customer cannot accept due to one of the reasons outlined in this item, neither has a legal responsibility to do so.

9.2 Non-force Majeure items

This limits the provider's ability to withhold delivery simply due to the provider not being able to obtain supplies at a cost which allows a profit to be realized. Force Majeure applies only to circumstances over which neither party has any control, as detailed in 9.1.

10. Notices by or to either provider or customer

This section outlines the acceptable methods or forms of communication between the provider and customer. The information would only be used if either party wanted to amend or change provisions during the contract term. Specific names and addresses of individuals are designated in the section to be used in correspondence by either party.

11. Miscellaneous provisions

11.1 Unenforceability of contract provisions

If any section of the contract is, or becomes, legally invalid or unenforceable, it does not negate the remainder of the contract. This common contract clause protects both parties against unenforceable conditions.

11.2 Benefit of this contract

This details the intended beneficiaries of the contract.

***11.3 Best-efforts clause**

This is typical of all self-help, direct purchase contract(s). It says that the provider will use best efforts to obtain natural gas but will not give a guarantee to do so. For this reason, it is important to investigate the provider prior to making any commitments.

Although this provision discourages many first-time purchasers, it must be remembered that even LDCs in their most expensive category of natural gas have been known to curtail customers. The simple fact is that if either a LDC or a non-LDC provider cannot obtain natural gas, it will not be delivered. By choosing a provider with multiple pipeline connections, the likelihood of non-deliverability is minimal.

11.4 Joint and several obligations of the customer(s)

All customers who are party to the contract are equally liable. For example, if two companies were to purchase and use natural gas under this contract, either would be liable in the event of non-performance (non-payment) by the other.

11.5 Entirety of contract

This item states that the contract as signed by the provider and customer is complete as written with no unattached side clauses or contracts.

11.6 Binding of parties to the contract

This item protects both provider and customer in the event their companies are sold or absorbed by another entity. If this happens, the successor company is obligated by the terms of the contract to the same extent as the original party.

11.7 Contract validity clause

This states that the contract was established subject to current rules by all governmental bodies having jurisdiction over contracts of this type. If a governmental body enacts legislation that changes or restricts provisions in the contract, then the legislated changes supersede the contract provisions.

11.8 Quality

This describes the natural gas as being pipeline quality, the generic term for all natural gas transported on any interstate pipeline.

11.9 Provider

Here is described the legal status of the customer/provider contract. The provider is a legal extension of the customer and can legally bind the customer to various contracts.

11.10 Customer credit

The provider can examine the creditworthiness of the customer at the inception of and during the contract period.

11.11 Taxes and Royalties

Here are described the various taxes and royalties, if applicable, that are paid by the provider in the customer's behalf up to the LDC city gate receipt point.

11.12 Measurement

Here is described the natural gas measurement process. Typically, natural gas is measured thermally (MMBtu/Dth) at all points up to the LDC city gate receipt point.

11.13 Transportation

The transportation of the natural gas from wellhead to LDC city gate receipt point is included in the provider's billing to the customer.

11.14 Titles, warranties, indemnity

Described in this section are the various titles, warranties, and indemnities included in this contract—how they apply to and affect each party.

11.15 Disclaimer of warranties

All warranties are contained in the written text of this contract and any other representations outside of the contract stipulations are without legal standing.

11.16 Third-party beneficiaries and relationship of the parties

There are no third-party beneficiaries to this contract. The only legal entities to this contract are the entities on the signatories to the contract page.

11.17 Limitation of liability

This describes the legal liabilities of each party to the other.

11.18 Waiver

This describes the effect on either party in the event of non-enforcement of any provision of this contract and the effect that this may have on the remainder of the contract.

11.19 Amendment

Here is described the procedure that the provider is to follow to amend the contract in the event of customer's LDC changes in rules/regulations that could increase or change the provider's costs.

11.20 Termination

The provider procedure to terminate this contract in the event that the customer defaults on provider payments, or misrepresents its financial data or any material data to the provider is described here.

11.21 Governing law

This describes the state in which this contract will be legally interpreted will be established. The state utilized should be the customer's state of incorporation. The provider will generally want their state of incorporation but this should be negotiable.

11.22 Confidentiality

Here are described the terms and conditions that apply to this contract relating to its confidential nature between the parties.

Signatories to the Contract

This is the page that when signed and dated by the proper provider and customer representatives, binds both parties to the contract terms. The executed section should be dated no earlier than the expected signature dates by the provider and customer.

Provider representative and customer representative may be located in different geographic areas, which can cause different signature dates for each. It should be at least the same as or later than the last provider/customer signature date.

Acknowledgements Page

Both parties are identified by company and name, and the signatures of the responsible parties are notarized in the locations of the provider and customer. When this page is properly signed by both parties and notarized, the contract *becomes binding* upon the *provider* and the *customer*.

Exhibit 'A'

This identifies the point of use of the contracted natural gas and the intrastate LDC who will distribute and deliver the natural gas to the customer.

Exhibit 'B'

This identifies the points of delivery by the provider to the transporting pipeline. It also identifies the receiving pipeline that will deliver the natural gas to the LDC.

Note: If possible, require the provider to specify more than one pipeline inlet. With a single inlet, mechanical and/or weather conditions could render it inoperable and restrict the natural gas source. Multiple inlets reduce or eliminate this possibility.

Exhibit 'C'

This provision applies to both, Item 4, Option A or B. It requires the provider to list the charge per unit of natural gas delivered to the customer. This is important to the customer since the provider's charge affects the overall cost of the natural gas to the customer. Many natural gas providers do not like Exhibit 'C' but will provide the required information if the customer insists.

How to Assure a Contract that Ultimately Serves the Needs of the Customer

Questions to ask a potential natural gas provider and points to consider prior to the signing of a contract include the following:

1. Is title to the natural gas for the entire term of the contract?

2. How many entry points are available into the transporting pipeline (wellhead meter points)?

3. Has non-delivery of natural gas ever occurred; and, if so, why?

4. Obtain a list of current clients, preferably in the customer's intrastate LDC's service area, and check out the performance of the provider with these customers.

5. Always ask a provider to quote: 1) natural gas costs, and 2) the total point-of-use cost. This point-of-use cost (commonly called burner tip) is what is paid in total for the natural gas delivered. It includes natural gas, interstate pipeline transportation, and shrinkage; and, intrastate pipeline transportation and shrinkage. This is the burner tip cost to use in determining savings.

6. Find out if the provider currently transports any natural gas over the customer's intrastate LDC's pipelines. If they do, contact the LDC and ask for their evaluation of the provider.

7. Always obtain and compare quotes and terms from more than one provider to determine the best deal.

8. Do not be afraid to negotiate on contract terms or natural gas costs, since providers are willing to do this if the requests are reasonable.

After reviewing these eight steps, find a provider who is honest and trustworthy. Selecting the right provider is the most important step in the process. It can only be done with knowledge of the process and reasonable expectations on the part of the customer.

THE RFP PROCESS

How to Have a Provider Supply the Contract

Once the definition of a customer-friendly contract is understood, the next step is approaching potential providers (brokers, marketers, and/or producers) to supply the information needed. One of the best ways is to supply each provider with an RFP similar to what vendors are required to submit for their products or services. It is very important that all prospective providers supply uniform information to the customer. If appropriate information is not provided, accurate correlation of data from the various quotes will not be possible.

A sample *Natural Gas Purchase RFP* is shown and can be utilized for both intrastate natural gas (Item #3, Pricing), as well as interstate natural gas (Item #4, Pricing). It should be utilized as a guideline only. It can and should be individually structured to a specific company's procedural requirements and needs.

All of these negotiation basics required to reduce natural gas costs, offer the potential for great savings. The process for realizing these savings is not difficult if the process is followed. It is always important to establish a working relationship with the LDC representative. Many times a special LDC program can be worked out that will save money for the customer.

Figure 17-2. The Natural Gas Project

REQUEST FOR PROPOSAL

Develop and submit for review, a contract document with appropriate attachments, to supply transportation natural gas to:

____
____
____

Use format and included all applicable information listed in attached *Guideline Specifications for Transportation Gas Contract*

All contract proposals must be submitted on or before:

________*(day)* __________*(month),* _________ *(year)*

Please send all proposals to:

____
____
____

Attention: ___________________________

Telephone: ___________________________

E-mail: ___________________________

Page 1

Figure 17-2. (*Continued*)

GUIDELINE SPECIFICATIONS FOR NATURAL GAS CONTRACT

1. Type of Contract

a) Spot market ______________________________

b) Fixed pricing – 6-months ______________________________

c) Fixed pricing – 12-months______________________________

d) Other structures, durations, etc. ______________________________

2. Nomination/Balancing Responsibility

a) Nomination: Provider________________Customer________________

b) Balancing: Provider________________Customer________________

3. Purchase Contract *(if applicable)*

a) Maximum monthly customer commitment____________________per Mcf/Dth

b) Minimum monthly customer commitment____________________per Mcf/Dth

***4. Pricing** *(For **INTRASTATE** natural gas only)*

a) Wellhead price $____________________per Mcf/Dth

_____Spot Market (Pricing Guide)

_____Fixed Pricing (for Term of Contract)

b) Provider's margin/fees/charges $____________________per Mcf/Dth

c) Shrinkage amount/cost (Intrastate) $____________________per Mcf/Dth

d) Miscellaneous Intrastate charges $____________________per Mcf/Dth

(List separately – e.g., thermal to volumetric conversion, monthly customer charges, etc.)

e) LDC transportation rate $____________________per Mcf/Dth

____

TOTAL COST AT BURNER TIP $____________________per Mcf/Dth

* Item #4 is utilized only if the natural gas commodity is physically located in the state where the customer's burner tip use point is physically located, intrastate *(within the state)*. States where this might be applicable would be Texas, Louisiana, Oklahoma, parts of Ohio, West Virginia, Pennsylvania, etc.

Page 2

Figure 17-2. (*Continued*)

GUIDELINE SPECIFICATIONS FOR NATURAL GAS CONTRACT

***5. Pricing** *(For __INTERSTATE__ natural gas only)*

Itemize as follows:

a) Wellhead price $__________ per Mcf/Dth

______Spot Market (Pricing Guide)

______Fixed Pricing (for Term of Contract)

b) Provider's margin/fees/charges $__________ per Mcf/Dth

c) Interstate transportation charges $__________ per Mcf/Dth

d) Shrinkage amount/cost *(__Interstate__)* $__________ per Mcf/Dth

e) Miscellaneous Intrastate charges $__________ per Mcf/Dth

(List separately – e.g., thermal to volumetric conversion, etc.)

f) LDC transportation rate $__________ per Mcf/Dth

g) Shrinkage amount/cost (Intrastate) $__________ per Mcf/Dth

h) Miscellaneous Intrastate charges $__________ per Mcf/Dth

(List separately – e.g., thermal to volumetric conversion, monthly customer charges, etc.)

TOTAL COST AT BURNER TIP $__________ per Mcf/Dth

* Item #5 is utilized only if the natural gas commodity is physically located in a state other than where the customer's burner tip use point is physically located. This type of arrangement is the most typical one utilized since most customer's burner-tip use points are not in the same state as the natural gas commodity – interstate *(between states)*.

Page 3

Figure 17-2. (*Continued*)

GUIDELINE SPECIFICATIONS FOR NATURAL GAS CONTRACT

5. Definitions

a) Lead time from execution of all contracts to first day delivery of natural gas.

b) Terms and conditions relating to billing and payments.

c) List of contracts negotiated in customer's behalf – wellhead, interstate, intrastate, etc.

d) Title to natural gas *(Where does it pass to the customer?)*

e) Force Majeure *(Specify conditions)*

f) Customer's remedy due to provider's error in regard to:

  (1) Nominations (Under/Over)

  (2) Banking *(Excess bank amounts on Local Distribution Company system)*

  (3) Balancing *(Interstate vs. Intrastate)*

g) Terms and payments of penalties for Items *(f-1, f-2, and f-3)*

h) Procedures and obligations for nominations and verification of nominations *(Customers and Marketers)*

6. Attachments

a) Source of natural gas *(Specifically identify)*

  (1) Location of natural gas – wellhead, gathering point, etc.

  (2) Specific contractual contracts, at/with wellhead, gathering points, interstate and intrastate pipelines, etc. in customer's behalf

7. References *(List using following guidelines)*

a) Current clients with this customer's Local Distribution Company *(LDC)*

b) Current clients with same general usage characteristics as this customer

c) Current clients that have been with provider for at least a one *(1)* year period

Page 4

SYNOPSIS—THE NATURAL GAS CONTRACT PROCESS

This chapter contained samples of a typical *Natural Gas Contract and Request for Proposal*. If the information contained in these items is understood and utilized, natural gas costs will be what they should be.

The natural as customer is responsible for their natural as costs. It is important, as a natural gas customer, to know how the process works in order to receive the benefits of natural gas cost reductions.

Chapter 18

The Monthly Natural Gas Billing Procedure

THE MONTHLY BILLING PROCEDURE

The Process

After the natural gas contract is signed and natural gas is flowing, monthly billings begin to arrive. Depending upon the arrangement, the different bills will generally be received from:

1. The natural gas provider (broker, marketer, producer)
2. The local distribution company (LDC)

When these bills first start arriving, it may seem that provider-supplied natural gas is not worth the effort. This chapter will be helpful in understanding the structure and meaning of these bills. Consequently, the process is not that complex if the terminology and use are understood.

The monthly billings will probably all arrive at different times. Therefore, when these bills are received and paid, it is wise to retain a copy of each so that a record of what was purchased and transported is available.

The bills are generally self-explanatory. With a little patience when first starting, the provider natural gas process will reap many dividends. A detailed discussion of the bills will be in the following sequence

1. The provider billing process
2. The local distribution company (LDC) billing process

The Provider Billing Process

The natural gas provider (broker, marketer, and/or producer) billing will include the natural gas commodity and interstate pipeline charges. This billing will be separate from the LDC billings. Most customer/

provider arrangements include both natural gas commodity and interstate pipeline transportation billing processes.

The pipeline transportation part of the billing will be included with the natural gas provider billing. This part of the billing consists of interstate transportation charges based upon the quantity of natural gas commodity delivered to the interstate pipeline by the natural gas provider.

For example, if the natural gas provider delivers 1,000 Dth to the pipeline, and the pipeline loss is 1 %, then 990 Dth would be delivered to the LDC, but the billing for interstate transportation will be for 1,000 Dth. The monthly billing will include the total amount of natural gas delivered whether it is consumed or banked.

Local Distribution Company (LDC) Billing Process

The local distribution company's billing process will remain much the same in the provider natural gas process. In the case of provider natural gas, the total quantity of natural gas transported by the customer, and the charge for this transportation and related items, will be detailed in the LDC's billing process.

It should be noted that the quantity of natural gas delivered to the customer may be different than the quantity of natural gas delivered to the LDC. The LDC's transportation charges are billed only when the natural gas passes through the customer's meter.

For example, the pipeline delivers 1,000 Dth to the LDC. The LDC delivers 800 Dth to the customer and the remaining 200 Dth (LDC shrinkage not included) is credited to the customer's bank. The customer's delivery charge will be for 800 Dth. The remaining 200 Dth will be billed when it is delivered to the customer.

In this example, the pipeline would charge for transporting 1,000 Dth (pipeline shrinkage not considered) and the LDC would charge for transporting 800 Dth (LDC shrinkage not considered).

The Nomination Process

Typically, the provider will perform the nomination process for the customer. However, the process remains the responsibility of the customer regardless of who actually performs the nomination process.

The nomination process of purchasing natural gas through other than LDC sources is termed third party (self-help). The expression self-help is appropriate because when purchases are made in this manner, the LDC no longer assumes or has responsibility for usages or the ordering of the natural gas supply. As a natural gas purchaser, the customer now assumes the responsibility for making these estimates of usage for the following month—or as it is termed nomination.

Following is an overview of what is normally required in the nomination process and an outline of a nomination outline. Due to widely varying policies and procedures of LDCs and pipelines, this particular form may not exactly fit every customer's needs, but it does contain the information normally required by any LDC or pipeline.

NOTE: Regardless of who actually performs the nomination process, the responsibility for the process remains with the customer.

Before a nomination is processed, some key information is required. This information should be about the customer's present operation, the LDC, and possibly the interstate transporting pipeline. The proper nomination depends upon an accurate estimate of natural gas consumption for the upcoming month.

Nomination Outline Details

To make this estimate, the following items must be considered. It is important that all of these items be completed by the provider, but the customer needs to be involved and know the process in order to make it work.

Item #1—Type of Usage

Consider what the usage for natural gas is: heating, cooling processing, or a combination of all of these items. For heating/cooling loads, consult past records for the same months. For processing loads, consider the current usage adjusting for any planned production schedule changes.

Item #2—Natural Gas Usage—Imbalance

To check for an imbalance, consider the actual usages in past months compared to what was nominated or ordered. Normally, the monthly LDC bill will indicate actual usages together with any imbalances that exist.

Generally, a bank or surplus of 1 to 5% over and above normal usage is good to have on hand. Most LDCs will not charge a penalty if this quantity is not exceeded. If the customer's bank is maintained, the nomination process is easier since the amount nominated will not have to be exactly what is used in any given month.

In addition, if the majority of the load is heating and/or air conditioning, using a similar month in a past year may not be an accurate guide as to what will be needed in a current month since weather changes can measurably affect usage rates. Generally, it is better to have a bank or surplus of natural gas rather than a deficit since any shortfall will be supplemented with normal LDC firm rate natural gas with the possibility of a penalty being charged.

Before resolving what size bank to establish, check with the LDC to determine their requirements and penalty conditions. In most instances, next month's natural gas needs will be nominated prior to receiving the current month's usage figures. So it will be necessary to either, estimate and use the bank to level out any inaccuracies in the estimating process; or, actually read the natural gas meter figures on a monthly basis.

Normally, using the same company personnel to process the nomination information each month will enable the estimate to be accurate enough if a small bank is maintained. If the natural gas meter(s) are actually read by this same person, then it is important to do the meter reading at about the same time as the LDC, making sure to include all meters since many facilities have more than one metering location.

NOTE: In actual application, most providers remotely read and/or monitor the customer's burner-tip meter data on a continuing basis electronically via a dedicated telephone line. However, the customer needs to understand the process of nomination banking and meter reading since it is the customer who is the responsible party for the entire process.

Item #3—The Policies of the LDC

The next areas to investigate are the LDC's policies, as follows:

<u>Balancing Policies</u>

The following questions will need to be answered:

a) What is the maximum bank or imbalance the LDC will allow?
b) How much time is allowed to adjust an imbalance?
c) Are there any charges or penalties for imbalance? If so, what are they?

<u>Fuel Shrinkage</u>

This refers to that portion of the natural gas that is delivered, which is retained or consumed by the LDC, to move the natural gas through the pipeline. Included in this shrinkage is:

a) Pumping station losses

b) Natural gas losses due to leaks
c) Unaccounted for natural gas losses

To compensate for this loss or shrinkage, slightly more natural gas will need to be ordered than will actually be used by the customer. A typical shrinkage factor for a LDC might be 1 to 5% of the total volume ordered. To compensate for this reduction, increase the amount needed by the shrinkage factor to arrive at what would be required to flow into the LDC's pipeline. Example

- Needed quantity—500 Dth
- Shrinkage factor—2%
- Ordered quantity with shrinkage included:
  (500 ÷ .98 = 526.3 or 527 Dth)

This natural gas shrinkage will have to be calculated as shown if the LDC requires nominations based upon the amount of natural gas received from the pipeline. If the LDC allows the nominations to be made for the amount of natural gas delivered to the customer, then the shrinkage calculation will not be required.

<u>Measurement Conversions</u>

Some conversions may be necessary if the LDC measures volumetrically (Ccf—100 cu. ft.) or (Mcf—1,000 cu. ft.) because providers sell by heating value—therm (100,000 Btu), dekatherm (Dth) (1,000,000 Btu), or MMBtu (1,000,000 Btu). The approximate conversions are as follows for changing from volume to heating value measurements (typically):

(1)-Ccf (100 cu. ft.) = (1) Therm (100,000 Btu) (±3%)
(1)-Mcf (1,000 cu. ft.) = (1) Dekatherm or MMBtu (1,000,000 Btu) (±3%)

These figures are approximate since all natural gas varies slightly in heat content. If this conversion is required, the LDC will provide the conversion factors if different from the approximations shown.

Policies of the Pipeline Company

In general, the same areas as were outlined in Item #3—The Policies of the LDC—will have to be investigated in the same manner for the Policies of the Pipeline Company.

When a customer is selecting a provider, it is important to consider the provider's ability to nominate, balance, and be responsible for errors that can cause customer penalties.

A low price is good only if it includes the customer services required to make the continuing process smooth and not require the customer to spend inordinate amounts of time fixing provider-caused problems. It is the customer who chooses the provider; and, it is the customer's responsibility to know the processes.

Natural Gas Nomination Process

A customer purchasing natural gas through a provider needs to be aware of the natural gas nomination process even though the provider usually takes responsibility for this function. The following steps have to be accomplished for the nomination process to be accurate.

1. Nomination date/Date nomination due
2. Nomination month
3. Days in nomination month
4. Expected usage
5. Desired bank
6. Current bank
7. Amount needed (Dth/Mcf)
8. Amount needed dally (Dth/Mcf)
9. LDC shrinkage factor (Dth/Mcf)
10. Conversion factor—Dth to Mcf
11. Pipeline shrinkage factor

Explanation—Natural Gas Nomination Outline Items

1. *Nomination date/Date nomination due*
   This entry indicates the nomination date the form is being completed and the date the nomination is due. Typically, the nomination due date is approximately two weeks prior to the date the natural gas is scheduled to flow.

2. *Nomination month*
   The month the natural gas is scheduled to be used.

3. *Days in nomination month*
   The number of days in the actual month the natural gas will be used.

4. *Expected usage*
   This is the expected natural gas usage during the nomination month.

5. *Desired bank*
   This number (Dth/Mcf) is the maximum quantity of natural gas that the LDC will allow a customer to accumulate, store, or have in reserve above what the customer actually intends to use during the nomination month.

   Typically, a LDC will allow a float or reserve in the range of 1 % to 5% above what the customer will actually use. The customer should always maximize the amount of natural gas in this category since it provides a cushion in case there are variations in the customer's nomination and actual usage.

6. *Current bank*
   This figure (Dth/Mcf) indicates the status of the bank carryover (if any) from the previous month. This number is important since it will be used to determine the beginning bank level for the nomination month.

7. *Amount needed*
   This amount (Dth/Mcf) will determine the total amount of natural gas to be nominated when factoring in current and desired bank levels.

8 *Amount needed daily*
   This number (Dth/Mcf) (if applicable) will be the total quantity of natural gas required on a daily basis including bank replenishment or reduction.

9. *LDC shrinkage factor*
   Most LDCs have a shrinkage or loss factor that must be calculated into the customer's total natural gas usage quantity. If there is a LDC factor, it is calculated here and added to the natural gas quantity shown in Item #8.

10. *Conversion factor—Dth to Mcf*
    All natural gas is purchased from the provider and delivered to

the LDC receipt point thermally. If the serving LDC measures volumetrically, a conversion must be completed (Ccf/Mcf).

11. *Pipeline shrinkage factor*
    Pipeline shrinkage factor considers the actual loss of natural gas as it travels through the interstate pipeline system. This loss is due to pipeline leaks and/or pressure boost station engine pumping losses.

The Monthly Follow-up

Once the provider natural gas is flowing, a monthly follow-up on the actual dollar savings on a monthly, as well as an accumulative basis should be completed. Following is a listing of items that should be on any provider natural gas monthly follow-up form. This outline will also provide the documentation to satisfy most company's accounting or cost -control requirements.

1. Nomination date
2. Documenting follow-up month
3. Days in follow-up month
4. LDC cost of natural gas
5. Quantity of natural gas purchased during month being analyzed on a daily basis
6. Quantity of natural gas purchased for entire follow-up month
7. LDC intrastate transportation charge
8. Interstate pipeline transportation charge
9. Natural gas cost
10. Provider charge
11. Miscellaneous charges
12. Total cost of provider natural gas for period being analyzed
13. Total cost of LDC natural gas if used for follow-up month
14. Monthly savings using provider natural gas
15. Accumulative savings

Explanation—Natural Gas Follow-up Outline Items

1. *Nomination date*
   The date the natural gas is nominated

2. *Documenting follow-up month*
   The month being analyzed—LDC billing period

3. *Days in follow-up month*
   The total number of days in the month being analyzed

4. *LDC cost of natural gas*
   This is the per-unit total cost of natural gas if the customer would have purchased natural gas through the LDC and not utilized a provider.

5. *Quantity of natural gas purchased during month being analyzed on a daily basis*
   This amount is the total quantity of natural gas at the wellhead meter point that the provider had to purchase for the customer. This quantity of natural gas will be greater than the customer burner-tip meter point quantity of natural gas actually used because of the various pipeline shrinkage factors and conversions in natural gas measurements (Ccf, Mcf, therm, Dth, MMBtu).

6. *Quantity of natural gas purchased for entire follow-up month*
   This amount (Dth per month) is simply the change from daily quantities in Item #5 to monthly quantities.

7. *LDC intrastate transportation charge*
   This represents the charge to move the customer's natural gas through the LDC's pipeline system. When the customer purchases natural gas directly from the LDC, this cost is included in the basic total rate cost for natural gas at the customer's burner-tip meter point. When a customer utilizes a provider, this cost is a separate item; however, when the customer purchases from the LDC directly, this item is not a separate item.

8. *Interstate pipeline transportation charge*
   The interstate pipeline transportation charge is similar to the one shown in Item #7 except that it is for the interstate pipeline portion of the customer natural gas transportation cost. As in Item #7, this charge is buried in total rate if the customer purchases natural gas directly from a LDC.

9. *Natural gas cost*
   This figure represents the provider's charge to the customer for

natural gas at the wellhead meter point and represents only the natural gas commodity cost. There are no costs to move the natural gas from the wellhead meter point to the customer's burner-tip meter point in this cost.

10. *Provider charge*

    This amount is the provider's charges to the customer for the services provided. The provider assists the customer in at least the following areas:

    a) Locates and provides the natural gas commodity at a reasonable cost
    b) Provides the customer with an efficient and economical interstate/intrastate pipeline routing for the natural gas to flow through
    c) Does all of the contractual work between the customer and interstate/intrastate entities
    d) Guarantees to the customer that natural gas flow will not be interrupted (assuming that the customer has opted for firm natural gas service) between wellhead and LDC city gate meter points, except for *force majeure* conditions
    e) Provides natural gas nomination and follow-up functions

    It is important to note, the provider generally cannot guarantee movement of a customer's natural gas through the LDC pipeline network (intrastate). However, it should be the provider's responsibility to assist a customer in negotiating a cost-effective firm natural gas transportation agreement with the customer's LDC.

11. *Miscellaneous charges*

    a) Natural gas storage fees apply to the customer purchasing natural gas and putting it into storage to be used at a later date. Typically, there are fees for injection, monthly storage, and extraction associated with this type of arrangement.
    b) Natural Gas Research Institute fees (GRI) can be assessed by an interstate pipeline to fund research into natural gas efficiency and new-use strategies. The Federal Energy Regulatory Agency (FERC) has to approve which interstate pipelines can assess these fees.
    c) Miscellaneous fees and/or charges assessed by the interstate pipeline.

12. *Total cost of provider natural gas for period being analyzed*
    This sum represents the total natural gas cost at the customer's burner-tip meter point for the total period being analyzed.

13. *Total cost of LDC natural gas if used for follow-up month*
    This figure represents what the total natural gas cost would have been if the customer would have purchased the entire month's natural gas through the LDC at the prevailing rates. This figure is important since it illustrates the cost that serves as the basis for savings calculations in Item #15.

14. *Monthly savings using provider natural gas*
    This amount is the cost differential (savings) between the LDC and provider natural gas.

15. *Accumulative savings*
    The amount from Item #14, plus last period's accumulative savings, equals this period's accumulative cost differential (savings) between the LDC and the provider natural gas costs.

SYNOPSIS—THE MONTHLY NATURAL GAS BILLING PROCEDURE

The Billing Process

It is more complex when utilizing a provider.

The Nomination Process

If these procedures are not understood, knowledge of true savings potentials will not be available.

This chapter provides information to fine-tune the natural gas cost reduction process. Provider natural gas purchase strategies cannot be successful unless a thorough knowledge of the billing process is established. Details, if not understood, can cause a good process to fail.

Chapter 19

Natural Gas Cost Reduction Strategies

At this point, all of the basics required to investigate the possibility of reducing natural gas costs have been discussed. The likelihood of being successful is directly proportional to the effort put into the process. The potential for savings is great and the process for realizing these savings is not difficult if the process is understood as outlined following:

1. Establish a working relationship with the LDC representative. Many times a special program can be worked out with the LDC that will save money and reduce the effort required in getting the program instituted.

2. If a program with the LDC is not possible, find out if the LDC offers firm or interruptible transportation of natural gas. If the LDC offers only interruptible transportation, calculate the cost of an alternate fuel source to determine net savings potential. Many times the alternate fuel source can be paid out of natural gas savings in less than one year.

3. Much care must be taken in the process of selecting a provider as a reliable relationship between customer and provider is the key to continuing satisfaction and natural gas savings. Do not neglect to ask questions and verify performance of a provider with current customers.

4. Take time each month to analyze and record the information on billings from the LDC and the provider since this data will be needed to accurately complete the nomination for the next month. It is important to record savings information on the monthly follow-up outline as shown in this publication.

The time to analyze the monthly bills and record savings information should take no more than one hour for each facility or location that is involved. The time required to do this monthly record keeping is a very small price to pay when compared to the savings which can be realized. Do not neglect this monthly process since an awareness of what is going on in the various parts of the program is very important.

5. If these steps are followed, a satisfactory, less costly natural gas program will result. Many times, savings of 5-20% are not unusual, and these are bottom-line savings.

   For example, a $30,000 annual natural gas cost savings in a company that earns 10% before tax profits means that $300,000 in new sales would have to be generated to provide the same $30,000 that was realized in the natural gas program.

Comparison of Natural Gas Costs with Electricity Costs

Frequently, natural gas and electricity compete with each other as energy sources for various uses such as heating, cooling, processing, etc. Both natural gas and electricity equipment providers tend to favor their own particular equipment to the exclusion of other types of processes.

It is important, as a natural gas or electricity customer, that a means of evaluating different natural gas and electric processes be available. When evaluating natural gas versus electricity for a process, one of the most important cost considerations is the energy cost. The only accurate way natural gas can be compared with electricity on a per unit cost basis is to arrive at a common denominator value that applies to both energy sources.

Since natural gas (Dth or Mcf) and electricity (kWh) are not comparable, another unit of measure must be used—the British thermal unit (Btu). If the Btu value of both natural gas and electricity are known, then an accurate comparison between the energy costs of the two processes can be determined. Shown in Figure 19-1 are the various factors that need to be known to accurately evaluate natural gas versus electricity energy costs.

Natural Gas/Electricity Cost Comparisons per 1,000,000 Btu

Since natural gas is generally priced in 1,000,000 Btu unit values (Dth/Mcf) and electricity is priced in kWh, it may seem that there is not an accurate way to compare natural gas to electricity costs. In Figure 19-2, a sample comparison is shown which illustrates the correct and accurate way to compare natural gas to electricity costs.

Since energy (natural gas/electricity) costs are only part of total costs, other items must be considered:

1. Equipment cost differences between natural gas and electricity.
2. Feasibility of utilizing natural gas and/or electricity in a given situation.
3. Burner efficiency of natural gas-fired equipment.

NATURAL GAS
*(1) Therm/Cef = 100,000 Btu *(1) Dth/Mef = 1,000,000 Btu
ELECTRICITY
(1) kWh = 3,412 Btu *(293.08) kWh = 1,000,000 Btu (1 Dth/Mcf)

*Thermal/volumetric measurements for the purpose of this chart are assumed to be equal. In practice, thermal/volumetric values can be as much as a 3% difference with relationship to Btu content.

Figure 19-1. British Thermal Unit (Btu) Comparison Chart

NATURAL GAS COST—	$10.50 (Dth/Mcf)
ELECTRICITY COST—	$0.07 (kWh)

Which energy source is less expensive?

ELECTRICITY—	$0.07/kWh	= $20.51/1,000,000 Btu
*NATURAL GAS—	$13.13/Dth	= $13.13/1,000,000Btu
Cost Difference		= $7.38 11,000,000 Btu

*At a burner efficiency of 80%, the cost of natural gas must be increased as follows: ($10.50 ÷ .80 = $13.13)

Figure 19-2. Sample of Natural Gas/Electricity Cost Comparisons

4. Supply reliability of natural gas versus electricity.
5. Electricity actual cost values on applications where heat pump processes can result in efficiencies of over 100% which reduces the true cost of electricity when compared to natural gas.

Miscellaneous liquid fuel conversions are shown in Figure 19-3. These conversions are necessary when trying to equate the cost of an alternate fuel to the cost of natural gas.

	FUEL OIL INFORMATION	
#2 Fuel Oil- (Average 140,000 Btu/gal)		
7.357 gallon	=	(1) Mcf natural gas (1,030,000 Btu + 140,000 Btu/gal)
7.143 gallon	=	1,000,000 Btu
#6 Fuel Oil- (Average 150,000 Btu/gal)		
6.867 gallon	=	(1) Mcf natural gas (1,030,000 Btu + 150,000 Btu/gal)
6.667 gallon	=	(1) Dth or (1) MMBtu or 1,000,000 Btu (1,000,000 Btu + 150,000 Btu/gal)
	PROPANE INFORMATION	
Average 91,500 Btu/gal		
11.257 gallon	=	(1) Mcf natural gas (1,030,000 Btu +91,500 Btu/gal)
10.929 gallon	=	(1) Dth or (1) MMBtu or 1,000,000 Btu (1,000,000 Btu +91,500 Btu/gal)
	BTU INFORMATION	
(1) Btu (British thermal unit)	=	The amount of energy required to raise the temperature of (1) pound of water; (1) degree Fahrenheit at temperatures between 58.5° and 59.2° Fahrenheit

Figure 19-3. Miscellaneous Liquid Fuel Conversion Data

What Customers Need to do to Reduce Natural Gas Costs

Become Proactive!

Doing nothing will increase your natural gas costs and risks.

What customers must have, understand, and manage

a) Have accurate natural gas usage data
b) Understand current status of natural gas deregulation
c) Understand individual facility natural gas usage characteristics
d) Manage natural gas commodity cost/usage

What customers need to develop, consider, and understand internally to control and reduce natural gas costs

a) Develop a natural gas strategy
b) Consider internal facility organizational factors
c) Understand commodity and natural gas contracts and their long-term implications

What Needs to be Done Now

Develop a natural gas cost control/reduction strategy

a) Know your current natural gas costs
b) Assess savings potentials
c) Utilize internal/external expertise to reduce natural gas costs

Items that should be included in any successful natural gas strategy

a) Usage characteristics:
    - Usage variables by day, month, season, etc.
    - Firm service
    - Interruptible service
    - Peak usage
    - Non-peak usage
b) Deregulated commodity purchasing
c) Rate schedule rate alternatives
d) Combined billing/metering if more than one meter utilized

SYNOPSIS—NATURAL GAS COST REDUCTION STRATEGIES

This chapter outlines natural gas costs in comparison to other energy sources and strategies to begin the cost reduction process.

Evaluating natural gas costs may not currently be a high priority in your company but it probably should be. The procedures described in this publication can become valuable tools in reducing natural gas costs. To accomplish goals, the right information is needed at the right time, in the right format.

In every company, whether for-profit or not-for-profit, there are probably areas of natural gas cost reduction opportunities available. In the scenario of today's natural gas cost instability, now is the time to get started with cost reduction strategies.

Natural gas costs are not going down and every day that a realistic cost reduction strategy is not utilized the opportunities lost will never be recovered. The process of natural gas cost reduction requires time and expertise, but delaying the process only reduces the savings potential and increases lost opportunities.

Section III

Electricity and Natural Gas

Chapter 20

Electricity and Natural Gas Hedging Strategies

HOW TO UTILIZE HEDGING IN THE ELECTRICITY AND NATURAL GAS PURCHASING PROCESS

The Strategy

Hedging is a strategy designed to reduce risk by utilizing forward and futures contracts, options, and swaps. Hedging can reduce volatility and reduce variability but rarely reduces total costs over time when compared to spot market pricing options.

The Process—Using Derivatives

Derivatives are specialized instruments that in and of themselves have no intrinsic value, but whose value is based on an underlying security. Derivative is a generic term that includes contracts, options, and swaps (forward and futures).

THE SPECIFICS—CONTRACTS, OPTIONS, SWAPS (FORWARD AND FUTURES)

Forward Contracts

A *forward contract* is a cash market transaction in which delivery of a commodity is deferred until after the contract has been made. It is not standardized and is not traded on organized exchanges. Although the delivery is made in the future, the price is determined at the initial trade date.

Futures Contracts

A *futures contract* is an agreement to purchase or sell a commodity for delivery in the future:

1. At a price that is determined at initiation of the contract
2. Which obligates each party to the contract to fulfill the contract at the specified price
3. Which is used to assume or shift price risk
4. Which may be satisfied by delivery or offset

A futures contract is an exchange traded, standardized agreement providing for the future exchange of a financial asset or commodity.

Options

Option contracts give their owner the right to buy a fixed amount of a specified underlying amount of natural gas at a fixed price at any time or on/before a fixed date.

Swaps

Swaps are agreements between at least two parties to exchange cash flows in the future according to a pre-specified formula. They can therefore be regarded as portfolios of forward contracts. The most common one is an agreement on the exchange of a fixed rate for a floating rate contract.

THE DETAILS

Forward Contracts

Forward contracts are a simple extension of cash or cash-and-carry transactions. Whereas in a standard cash transaction, the transfer of ownership and possession of the commodity occurs in the present delivery under a forward contract is delayed to the future.

For example, customers may enter into forward contracts to secure a commodity (electricity/natural gas) for future delivery. This avoids volatility in spot prices and the need for commodity storage for extended periods.

Forward contracts are varied, but they all tend to deal with the same aspects of a forward sale. All forward contracts specify the type, quality, and quantity of commodity to be delivered as well as when and where delivery will take place.

In addition, forward contracts set a price or pricing formula. The simplest forward contract sets a fixed (firm) price. More elaborate, price-setting mechanisms include floors, ceilings, and inflation escalators. By setting such a price, the buyer and seller are able to reduce or eliminate uncertainty with respect to the sale price of the commodity in the future.

Knowing such prices with certainty may allow forward contract users to better know their costs. Overall, forward contracts are designed to be flexible so as to match the needs of the parties entering into them.

Direct results of the forward pricing and delivery features of forward contracts are default and credit risks. In the case of long-term forward contracts, the exposure to default and credit risks may be substantial. To deal with the risk of default, parties must evaluate the creditworthiness of counterparties and deal only with parties that maintain good credit ratings. Ultimately, how parties deal with default and credit risk in a forward contract is up to them.

Futures Contracts

A futures contract obligates each party to buy or sell a specific amount of a commodity at a specified price. Unlike a forward contract, buyers and sellers of futures contracts deal with an exchange (NYMEX), not with each other.

For example, a producer wanting to sell a commodity in December can sell a futures contract through the New York Mercantile Exchange (NYMEX), and a buyer can buy a December commodity future from the exchange. The December futures price is the one that causes offers to sell to equal the bids to buy; i.e., the demand for futures equals the supply.

The December futures price is public, as well as the volume of trade. If the buyer of a December futures contract finds later that the commodity is unneeded, the buyer will have to sell a December commodity futures contract at the prevailing price in order to not be obligated of the original contract. Since both parties have bought and sold a December commodity futures contract, then each has met the obligations to the exchange by netting them out.

Several aspects of futures contracts are worth emphasizing. A buyer who elects to hold the contract until maturity is guaranteed the initial price paid for the contract. The buyer of the futures contract can always demand delivery; the seller can always insist on delivering.

As a result, at maturity the December futures price for the commodity and the spot market price will be the same. If the commodity price were lower, people would sell futures contracts and deliver the commodity for a guaranteed profit. If the commodity price were higher, people would buy futures and demand delivery, again for a guaranteed profit.

Only when the December futures price and the December spot price are the same, is the opportunity for a sure profit eliminated.

A trader can sell commodity futures (short position) even though there is no access to the commodity. Likewise, a trader can buy (long position) even though there is no use for the commodity.

Speculators routinely buy and sell futures contracts in anticipation of price changes. Instead of delivering or accepting the commodity, speculators close out positions before the contracts mature. Speculators perform the useful function of taking on the price risk that producers do not wish to bear.

Futures contracts allow a trader to make a commitment to buy or sell large amounts of a commodity for a very small initial commitment (margin). Consequently, traders can make large profits or suffer huge losses from small changes in the futures price. This leverage has a history of being the source of spectacular failures in the past.

Futures contracts may not, by themselves, be useful for all those who want to manage price risk. Futures contracts are available for only a few commodities and a few delivery locations. Also, they are not available for deliveries a decade or more into the future. However, there is a large business conducted outside of these exchanges in the over-the-counter (OTC) market to supplement futures contracts and better meet the needs of individual companies.

Options

An option is a contract that gives the buyer of the contract the right to buy (a call option) or sell (a put option) at a specified price (strike price) over a specified period of time.

American options allow the buyer to exercise the right to either buy or sell at any time until the option expires. Whether the option is sold on an exchange or on the OTC market, the buyer pays for the contract up front.

For example, the option to buy a one contract of a commodity at a price of $10.50 per unit in December may have a premium of $0.50 per unit. If the price in December exceeds $10.50 per unit, the buyer can exercise the option and buy the commodity for $10.50 per unit. More commonly, the entity offering the option pays the buyer the difference between the market price and the option price.

If the commodity price is less than $10.50 per unit, the buyer lets the option expire and loses $0.50 per unit. Options are used successfully to put floors and ceilings on prices; however, they tend to be expensive.

Swaps

Swaps (also called contracts for differences) are the most recent innovation in finance. Swaps are created, in part, to give price certainty at a cost that is lower than the cost of options.

A swap contract is an agreement between two parties to exchange a series of cash flows generated by underlying assets. No physical commodity is actually transferred between the buyer and seller. The contracts are entered into between the two parties or principals, outside any centralized trading facility or exchange, and are characterized as OTC derivatives.

Because swaps do not involve the actual transfer of any assets or principal amounts, a base must be established in order to determine the amounts that will periodically be swapped. This principal base is known as the notional amount of the contract.

For example, one person might want to swap the variable earnings on a million dollar stock portfolio for the fixed interest earned on a treasury bond of the same market value. The notional amount of this swap is $1 million. Swapping avoids the expense of selling the portfolio and buying the bond. It also permits the investor to retain any capital gains that the portfolio might realize.

Many of the benefits associated with swap contracts are similar to those associated with futures or options contracts. That is, swap contracts allow users to manage price exposure risk without having to take possession of the commodity. Swaps differ from exchange-traded futures and options in that, because swaps are individually negotiated instruments, users can customize them to suit the risk management activities to a greater degree than is easily accomplished with more standardized futures contracts or exchange-traded options.

Although swaps can be highly customized, the parties are exposed to high loss risk because the contracts generally are not guaranteed by a clearinghouse as are exchange-traded derivatives. In addition, customized swaps generally are less liquid instruments, usually requiring parties to renegotiate terms before prematurely terminating or offsetting a contract.

GUIDELINES FOR ASSESSING HEDGING ACTIVITIES

Establish the Need

Because hedging is a costly activity, having evidence that a customer is willing and able to pay for that service is important. For example, a customer may prefer catastrophic protection meaning protection from the possibility of extreme price spikes. If so, that may reveal a preference for a price cap approach (and thus, the use of options).

Always Keep the Hedging Strategy Simple

Commodity procurement decisions can be separated from hedging decisions. Customers can purchase the required commodities at index and then use financial derivatives to hedge the risks inherent in such a purchasing practice. It is probably easier and just as effective to pursue hedging-strategy objectives separate and distinct from the commodity procurement objectives. When hedging decisions become commingled with commodity purchase decisions (as in the case of fixed-price commodity contracts), it is more difficult to assess the prudence of that bundled decision.

From the perspective of customers, price risk can be defined as the product of the probability of high prices and the magnitude of losses from such prices; i.e., the harm done to customers when they pay extremely high commodity prices.

Traditionally, commodity procurement and price-risk management have been bundled as one activity; i.e., the case of forward contracts. The advent of financial instruments such as futures contracts, options, and swaps allows customers to procure the commodity and manage price risks as separate unbundled activities. As a feasible if not economical strategy, customers could purchase all of their physical commodities on the spot market and, separately purchase futures contracts to stabilize prices.

Overall, commodity procurement has more of a least-cost (or a so-called 'best cost' objective, whereas price-risk management has the effect of aligning the range of prices to better match the customer's tolerance for risk.

Specify the Objectives of a Hedging Strategy in Detail

It is important to identify the general objectives of the hedging strategies. This requires identifying the specific risks being managed, as well as the specific hedging tools that will be used. Also, it requires an explanation of the role, if any that price expectations play in the proposed hedging strategies. The anticipation of prices falling may cause the end result to be limited hedging activities. This could be defensible depending on the reasonableness of expectations.

Probably, the major goal of hedging is to limit extreme commodity price spikes during certain times of the year. By doing so, unacceptably high commodity bills can be avoided, which probably eliminates much of the economic harm from volatile commodity prices. *It is incorrect to expect hedging to reduce the average cost of purchased commodities over time.* Since a premium would be paid to shift risk, it would be expected that the average cost would be higher.

Identify All Hedging-strategy Costs

Needless to say, all costs, whether potential or actual, should be identified. Recovery provisions should be clearly articulated. It must be understood that hedging activities are costly, and it should be understood that expenditures on hedging do not always produce a benefit. Risk protection may be purchased and yet not be needed. For example, money can be spent on options that are never exercised. Or, by using futures, a price may be effectively locked in that turns out to be in error.

Understand Hedging Abilities

Hedging strategies should be designed and operated by sufficiently qualified personnel (probably not the customer). Temptation to micromanage the hedging strategy should be resisted. Instead, attention should be focused on the general provisions and parameters of the overall strategy.

Successful Hedging Strategies

The reasonableness of a hedging strategy should be understood before a program is actually implemented. If it is decided to perform after-the-fact reviews, then there will be the risk of creating unrealistic or inefficient performance standards, or both. The success of a hedging strategy should be evaluated not only on how things turn out, but also on whether the objection of the process was achieved.

Hedging Risks

In the commodity industry, risk management has become an integral activity of market participants. With unpredictable and volatile commodity prices, the industry has developed products transferring risks between parties, which have resulted in many having interest in the use of these products.

Hedging is one of those activities, similar to the purchasing of insurance, where by design it is expected to result in a *net loss*. Consequently, hedging is vulnerable to after-the-fact criticism. But, if its intent is to avoid large variations, a peace-of-mind type of benefit can still be regarded as successful and prudent.

SYNOPSIS—ELECTRICITY AND NATURAL GAS HEDGING STRATEGIES

Should hedging be used? It depends upon one basic consideration: Is price stability or variability more acceptable? Hedging strategies have one basic purpose— *price stability*.

If price stability for a period in the future is important, hedging may be the strategy to utilize. If least cost is important, hedging may not be the best strategy. What to do depends more upon corporate philosophies than with the process utilized.

Generally, price stability and lowest cost do not exist in the same strategy. Hedging is designed to reduce price variability, not necessarily long-term costs.

Chapter 21

Electricity and Natural Gas Savings Opportunities

PUTTING ALL OF THE INFORMATION TO USE

Now that all of the basic electricity and natural gas information has been covered, the next portion of the process involved is actually analyzing the facility where electricity and natural gas are used. Following is a questionnaire and/or evaluation outline that can be utilized in this effort. This outline is divided into the following four sections.

1. **General Description of the Facility**

This section consists of (14) general questions relating primarily to the physical attributes of the company as well as employee data. The reason for this generalized type of information is that the current natural gas usage characteristics, as well as the future characteristics, can be dependent upon facility specific considerations. This section data is important as it has considerable value in many areas of planning and forecasting in general.

2. **Rate Schedule Considerations**

This section consists of (3) questions that address the current rate schedules that serve the facility.

3. **Rate and Meter Data**

This section contains (5) questions concerning specific information about current rate schedules as well as meter data.

4. **General Usage Data**

This section contains (6) questions relating to general usage information that is necessary to be able to determine the impact usage characteristics have on overall electricity and natural gas costs.

How to Use the Outline

By properly utilizing this outline, the electricity and natural gas user will have a complete picture of all of the variables that can cause costs to be more than necessary, either now or in the future. As with any process, the evaluations of the costs are dependent upon relevant and accurate base data.

Some of the questions presented on this outline may at first seem to have little to do with electricity and natural gas usage or costs, but when all of the questions are put together, the rationale for them will be understood. These questions relate to the basic data required to accurately assess current uses and costs, as well as in developing future strategies to reduce costs.

There are no short cuts to success. Time utilized at the beginning of the analyzation process will pay many dividends in the accuracy of the information developed.

ELECTRICITY AND NATURAL GAS
QUESTIONNAIRE / EVALUATION OUTLINE

Section #1—Facility Description

1. Number of square feet in total facility
2. Number of square feet in office space
3. Number of square feet in non-office space
4. Number of employees per shift of operation (first, second, third)
5. Daily hours of operation
6. Number of days and hours worked per week
7. Plans for the facility for 1-, 2-, 3-years in the future for products, facility size, etc.
8. The approximate age of the facility.
9. Is the property leased, rented, or owned?
10. If the property is not owned, how many more years does the lease or rental agreement cover?
11. Are there any railroad sidings adjacent to the facility?
12. Is this a for-profit or not-for-profit facility?
13. Does the facility have an energy management system (EMS)?
14. Does the facility utilize any energy accounting procedures currently?

Section #2—Rate Schedule Considerations

1. Names and addresses of the serving electric and natural gas utility companies
2. Any communication with the utility service representatives
3. Are there a copies of rate schedules available

Section #3—Rate and Meter Data

1. Rate schedules utilized in the facility
2. Rate riders utilized
3. Number of electricity and natural gas meter points
4. Any problems with meters that have resulted in reading errors
5. Any questions or problems relating to meters, or metering in general

Section #4—General Usage Data

1. Usages of electricity and natural gas for this facility (heating, absorption air conditioning, manufacturing processing, heat-treating, or processing furnaces, etc.).

2. Quantify electricity and natural gas usage variations from month-to-month.

3. Potential for electricity and natural gas usage curtailments if an interruption of service were to occur.

4. Electricity and natural gas onsite backup or alternate fuel sources availability.

5. If electricity and natural gas are used in production processes, are there any opportunities to recover/utilize waste heat from the processes.

6. From general observations of the facility's a) usage patterns, b) hours of operation, or c) types of operations performed, are there any general comments that would be helpful in identifying areas of potential cost reduction?

SYNOPSIS— ELECTRICITY AND NATURAL GAS SAVINGS OPPORTUNITIES

This chapter describes the beginning of actually using the information that has been presented to this point. A disciplined, continuing electricity and natural gas cost reduction processes will result in many quantifiable savings opportunities. The most important part of any reduction process is getting started in a well-defined, disciplined process.

To realize the value of the material that has been presented in this publication, you need to do something. Knowledge and action go together to accomplish a goal. At this point, you have knowledge— the action part requires the utilization of this knowledge in some affirmative manner.

HAVING KNOWLEDGE WITHOUT ACTION ACCOMPLISHES VERY LITTLE.

BE PROACTIVE

USE WHAT YOU KNOW, TO DO WHAT YOU NEED TO DO.

Index